SpringerBriefs in Applied Sciences and Technology

T0212821

More information about this series at http://www.springer.com/series/8884

Felipe Richter Reis

Editor

Vacuum Drying
for Extending Food
Shelf-Life

Springer

Editor
Felipe Richter Reis
Instituto Federal do Paraná
Jacarezinho
Paraná
Brazil

ISSN 2191-530X   ISSN 2191-5318   (electronic)
ISBN 978-3-319-08206-6   ISBN 978-3-319-08207-3   (eBook)
DOI 10.1007/978-3-319-08207-3

Library of Congress Control Number: 2014944541

Springer Cham Heidelberg New York Dordrecht London

Printed on acid-free paper

Springer is part of Springer Science+Business Media (www.springer.com)

I would like to dedicate this book to my family, who always supported me, and to my wife, Daiany, for her unrestricted love

Preface

This book was written for presenting to researchers and engineers of the food area examples of studies where high quality dried food products were obtained by drying under subatmospheric conditions. Vacuum drying techniques are especially useful for heat-sensitive foods. The book is based on scientific articles published from the 1940s to the present, and is divided into five chapters: the first is an introduction to the use of vacuum in food processes; the second, third and fourth comprise advances on the most relevant vacuum drying techniques applied to food, namely conventional vacuum drying, freeze-drying and microwave-vacuum drying and the fifth deals with studies where vacuum drying techniques were compared to other drying techniques in terms of quality and/or engineering aspects. The book can be used as a reference for anyone who is interested in the use of vacuum drying for extending food shelf-life, whether it is for academic purposes or practical applications.

Acknowledgments

The authors thank the Brazilian Government as represented by the Coordenação de Aperfeiçoamento de Pessoal de Nível Superior—CAPES, for providing them with access to the Portal de Periódicos da Capes. This tool allows free access to Brazilian federal researchers to several scientific journals. Without the Portal de Periódicos da Capes it would not be possible to write this book.

Acknowledgements

Contents

Introduction to Low Pressure Processes

Felipe Richter Reis

Abstract This chapter deals with general aspects of food vacuum drying. Food drying under vacuum presents several advantages over conventional atmospheric drying. Vacuum drying takes place in the absence of oxygen and at mild temperature, thus preserving most of the food nutritive and sensory characteristics. Nevertheless, it is usually a more costly process when compared to atmospheric drying. Conventional vacuum drying, freeze-drying, microwave-vacuum drying and related processes are the most known vacuum drying processes. While conventional vacuum drying is the cheapest process, microwave-vacuum drying is the fastest and freeze-drying provides the product with the best quality. On the other hand, conventional vacuum drying yields products of inferior quality and freeze-drying is very expensive, while microwave-vacuum drying presents intermediate costs and provides the product with acceptable quality. Summarizing, each one of these processes presents strengths and weaknesses in terms of engineering and quality aspects. The choice for one or other vacuum drying technique will depend on the needs to be fulfilled.

Keywords Food drying · Vacuum drying · Microwave-vacuum drying · Freeze-drying

Drying is one of the oldest methods for extending food shelf life. In ancient times, the heat of the sun and the salt of the sea were used for drying food. It was not possible to control the drying conditions at that time. After centuries of development in drying techniques and equipment, men are now able to control the drying conditions, such as time, temperature and pressure. The use of reduced pressure for drying food is the focus of this book.

One food shelf-life is closely dependent on its water content. Nevertheless, differently from what would be expected, it is not the absolute moisture content of the food that defines its shelf-life. Free water is the water available for microbial,

F. Richter Reis (✉)
Food Technician Course, Instituto Federal do Paraná, Jacarezinho, Brazil
e-mail: felipe.reis@ifpr.edu.br

© The Author(s) 2014
F. Richter Reis (ed.), *Vacuum Drying for Extending Food Shelf-Life*, SpringerBriefs
in Applied Sciences and Technology, DOI 10.1007/978-3-319-08207-3_1

1

Table 1 Absolute pressure that must be attained in the drying chamber for boiling the water contained in the food at different drying temperatures [4]

Water boiling temperature (°C)	Absolute pressure (kPa)
20	2.3385
30	4.2461
40	7.3837
50	12.350
60	19.941
70	31.188
80	47.390
90	70.139

enzymatic and chemical reactions, the three main sources of deterioration in food. The free water content, named water activity is the property that regulates food shelf-life. Water activity can be defined as the vapor pressure of water in the food divided by the vapor pressure of pure water at the same temperature. Nowadays this property can be easily measured by digital water activity meters. The obtained numerical result is very useful for establishing the end of the drying process.

Drying under vacuum is more expensive than atmospheric drying, due to the fact that vacuum dryers need to be strong enough to withstand the pressure difference between the inside and the outside of the drying chamber. Thus, there must be advantages to justify the use of vacuum drying instead of atmospheric drying. Among these advantages, the greatest is the superior quality of the product which results from processing at lower temperature. The absence of oxygen under vacuum constitutes an additional advantage when drying oxygen-sensitive materials. Increased drying rates also contribute.

It is well-known that at atmospheric pressure, 101.325 kPa, water boils at 100 °C. During a vacuum drying process, subatmospheric pressures are achieved inside the drying chamber. Therefore, it is possible to boil the water contained in the food at lower temperatures when compared to atmospheric drying. Table 1 presents temperatures that are commonly used for vacuum drying of foods and the respective pressure that must be achieved to boil the water contained in the food.

When it comes to food drying, severe processes where high temperatures are attained usually destroy the food nutrients and functional compounds. For instance, the yacon (*Smallanthus sonchifolius*) is a functional vegetable that is native to the Andes, South America. Raw yacon is rich in phenolic compounds and dietary fibers. Nevertheless, yacon fibers and phenolics seem to be destroyed at 70 °C [10] and 80 °C [11], respectively. Vacuum drying of yacon, consequently, is a way to increase its shelf life, which is originally short, while preserving its health promoting properties [9].

The simplest equipment for vacuum drying of foods is a vacuum oven. In this equipment, a drying chamber containing a metallic plate at the bottom that is heated by an electrical resistance receives the food sample. The food is usually placed inside a glass or metal container as to form one single layer. A pump provides the chamber with vacuum. A gauge called vacuometer shows the value of the pressure inside the chamber. A digital display and buttons allow the temperature control. Figure 1 shows the schematic of a vacuum oven.

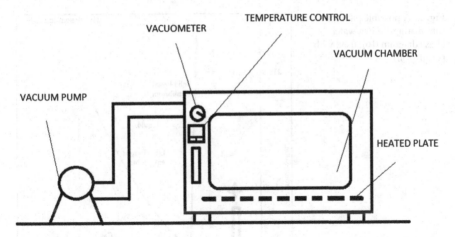

Fig. 1 Schematic of a vacuum oven

Besides the conventional vacuum drying, usually carried out in a vacuum oven, other processes were developed to dry foods under low pressure. Among them, freeze-drying and microwave-vacuum drying are the most studied. The former consists in removing water from the food by means of ice sublimation. The latter comprises the use of microwaves for heating the food inside a chamber under vacuum.

Sometimes, in order to obtain an acceptable product, it is necessary to use very low temperatures and higher vacuum, in a process originated approximately in the 1940s and named freeze-drying. This process provides food with several desirable features, such as: high stability to chemical changes, due to the use of temperatures below freezing point; low loss of aromatic volatiles, for the same reason; exclusion of foaming phenomena, due to the fact that the product to be dried is a frozen solid; permanent dispersion, i.e., absence of solute concentration as the frozen solvent sublimes; absence of case-hardening effects, i.e., no crust is formed in the product; maximum sterility, since the drying process is conducted under frozen conditions and the final product is "fully-dried" [2].

The process of freeze-drying could be described as follows: initially, the product is frozen solid; then it is exposed to a controlled temperature-pressure environment in a suitable chamber; the pressure in the chamber is regulated as to promote the direct transition from the solid state to the vapor state (sublimation), avoiding ice melting. When pressure is reduced below 0.006 atm, solid ice sublimates. For example, at 0.003947 atm, ice sublimates at −6 °C; at 0.000658 atm, ice sublimates at −25 °C. Although, the water in food is seldom pure, presenting soluble compounds that low its freezing point. Therefore, it is necessary to be aware of the eutectic point, which can be defined as the lowest freezing point of a food, being directly related to this food composition [5]. One possible path for the sublimation process of ice is shown by arrows in Fig. 2.

The typical conditions of freeze-drying are product surface temperature between 35 and 80 °C and chamber pressure between 13 and 270 Pa. Whichever

Fig. 2 A possible path for sublimating ice in a water phase diagram ([6]; used with permission)

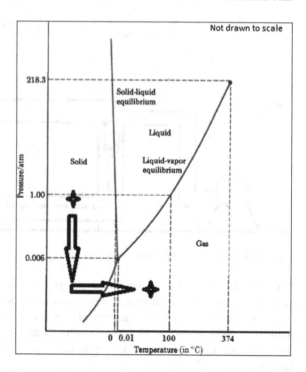

combination of temperature and pressure is used it is preferable that the food layer be thin in order to facilitate the heat and mass transfer. Regarding the characteristics of the equipment, a freeze-dryer comprises basically: a vacuum chamber, a heat source, a condenser and a vacuum pump. The vacuum chamber can be either cylindrical or rectangular. The heat source can be one or two hollow plates heated by hot water or electrical resistance, which might be combined with microwave heating [8]. Figure 3 shows the schematic of a freeze-dryer.

The other vacuum drying process of relevance is microwave-vacuum drying. The use of microwaves in drying improves the drying rates due to the porous structure created during water evaporation by dielectric energy, facilitating the mass transfer. While a typical convective drying usually takes hours, a typical microwave-vacuum drying takes minutes. On the other hand, the heat distribution in the material is not uniform [1].

As in other microwave equipment, in a microwave-vacuum dryer the heat is obtained by electromagnetic radiation. Such energy is absorbed by the water molecules contained in the food and converted into kinetic energy. This promotes an intense vibration in water molecules, generating heat, which is used for their evaporation. The results are high drying rates at low temperatures. Nevertheless, in order to obtain homogeneous sample heating it is necessary to use samples with small thickness [7].

A typical microwave-vacuum dryer of foods is composed by a microwave oven, a glass vacuum desiccator, a vacuum pump, a pressure regulator and a condenser. Figure 4 shows the schematic of a microwave-vacuum dryer.

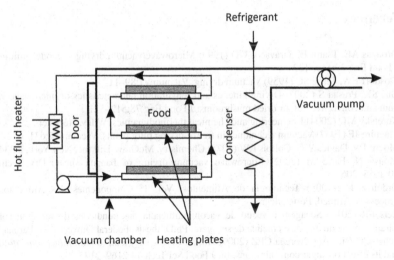

Fig. 3 Schematic of a freeze-dryer ([8], modified)

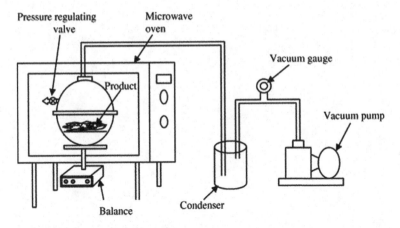

Fig. 4 Schematic of a microwave-vacuum dryer [3]

The procedure of drying in this apparatus consists in placing the food inside the vacuum desiccator, sealing it air-tight and applying the desired pressure and radiation. Weight loss can be measured either by breaking the vacuum and taking the sample to a balance or weighing the sample in loco, such as in Fig. 4. The most common way of operation is with intermittent radiation, in order to avoid material overheating.

Once the most relevant vacuum drying methods are sufficiently known, their previous applications to food will be dealt with in the following chapters.

References

1. Drouzas AE, Tsami E, Saravacos GD (1999) Microwave/vacuum drying of model fruit gels. J Food Eng 39:117–122
2. Flosdorf EW, Tease SC (1959) Vacuum drying. Vacuum VI:89–112
3. Giri SK, Prasad S (2007) Drying kinetics and rehydration characteristics of microwave-vacuum and convective hot-air dried mushrooms. J Food Eng 78:512–521
4. Koretsky MD (2004) Engineering and chemical thermodynamics. Wiley, Hoboken
5. Maguire JF (1967) Vacuum techniques in freeze-drying pilot plants. Vacuum 17:631–632
6. Moore JW, Davies WG, Collins RW (1978) Chemistry. McGraw-Hill Inc, New York, p 334
7. Mousa N, Farid M (2002) Microwave vacuum drying of banana slices. Dry Technol 20:2055–2066
8. Ordóñez JA (2005) Tecnologia de Alimentos: Vol. 1: Componentes dos Alimentos e Processos. Artmed, Porto Alegre
9. Reis FR (2011) Secagem a vácuo de yacon: influência das condições de secagem sobre parâmetros de qualidade e cinética de secagem. Ph.D. thesis, Federal University of Paraná
10. Scher CF, Rios AO, Noreña CPZ (2009) Hot air drying of yacon (*Smallanthus sonchifolius*) and its effect on sugar concentrations. Int J Food Sci Tech 44:2169–2175
11. Takenaka M, Nanayama K, Ono H et al (2006) Changes in the concentration of phenolic compounds during growing, storing, and processing of yacon. J Jpn Soc Food Sci 53:603–611

Studies on Conventional Vacuum Drying of Foods

Felipe Richter Reis

Abstract This chapter deals with the simplest way to perform the vacuum drying of foods. Such process is usually performed as a batch in a vacuum oven. The chapter presents studies carried out from year 2001 to year 2014 as surveyed in electronic databases. The vacuum dried foods used as material in these studies include herbs, fruits, vegetables and mushrooms. Most of these foods present health-promoting (functional) properties which are preserved even after drying due to the special conditions (low heating and low oxygen content) used during vacuum drying. Effective moisture diffusivity, a property related to the ability of moisture to migrate from the product to the environment, was calculated in many of these studies and the respective values are presented here. An equation for calculating this parameter in foods is also presented. Several thin-layer drying models, which are equations that fit the values of moisture content during drying, are also presented.

Keywords Vacuum drying · Effective moisture diffusivity · Drying models

In the beginning of the last decade, the vacuum drying of celery slices was optimized toward final product quality by means of the response surface methodology [17]. The effect of drying temperature (65–75 °C), slice thickness (1–3 mm) and pressure (16–20 inHg) on the rehydration capacity, bulk density, moisture content and overall acceptability of the final product was evaluated. When significant effects of the variables on the responses were observed, they were fitted by using polynomial models. All of the quality parameters studied, except from the sensory scores, were well predicted by using the proposed models. After plotting all possible combinations of variables in response surface plots, the graphs were superimposed in order to yield optimum drying conditions. The optimum drying conditions were: drying

F. Richter Reis (✉)
Food Technician Course, Instituto Federal do Paraná, Jacarezinho, Brazil
e-mail: felipe.reis@ifpr.edu.br

F. Richter Reis (ed.), *Vacuum Drying for Extending Food Shelf-Life*, SpringerBriefs in Applied Sciences and Technology, DOI 10.1007/978-3-319-08207-3_2

temperature of 74.5 °C, vacuum pressure 17.5 inHg and slice thickness of 1 mm. When using the optimum conditions, a final product presenting high rehydration capacity, low bulk density, low moisture content and probably high overall acceptability was obtained.

The effect of different drying conditions and modified mango pulp composition on drying time, color and effective moisture diffusivity was evaluated by Jaya and Das [10]. The variables tested were initial thickness of pulp (2–4 mm) and vacuum chamber plate temperature, i.e., drying temperature (65–75 °C). Drying pressure was kept constant at 30–50 mmHg. The drying kinetics was adequately fitted by a model based on effective moisture diffusivity. The addition of tri calcium phosphate, maltodextrin and glycerol monostearate to the mango pulp led to lower values of effective moisture diffusivity. However, additives were necessary for allowing the grinding of pulp into a powder. Increase in pulp thickness and drying temperature led to increase in effective moisture diffusivity. Increased drying rates were associated with low product thickness and high temperatures. With regard to color, the reconstituted pulp powder was used as sample. The higher the product layer and the drying temperature, the higher the total color changes. In conclusion, it was recommended the use of a product layer lower than 2.6 mm and a drying temperature lower than 72.3 °C for obtaining high quality dried mango pulp.

Zhang et al. [32] optimized low-vacuum (0.67 kPa) drying conditions for obtaining dried sweet pepper with high contents of Selenium and chlorophyll. A three factor-three level response surface design with three replications at the center point was used to estimate the effects of drying temperature (60–100 °C), blanching time (3–13 min) and blanching solution pH (1–11) on the levels of Selenium and chlorophyll in the final product. Blanching solution pH was the factor that affected the most the Selenium content, while drying temperature markedly affected the chlorophyll content. By using contour plots and differentiating the second order polynomial regression equation, the authors observed that the optimum drying conditions were a dying temperature of 75 °C, a blanching solution pH of 7.0 and a blanching time of 8 min. When the optimum process conditions were used, maximum levels of Selenium (191.2 µg/g) and chlorophyll (187.8 µg/g) were observed in the dried sweet peppers.

Arévalo-Pinedo and Murr [2] modeled the vacuum drying kinetics of pumpkin slabs under various drying conditions, studied the impact of blanching and freezing on the drying rates and evaluated the impact of shrinkage on the effective moisture diffusivity of the slabs. The drying temperature varied from 50 to 70 °C and the absolute pressure in the drying chamber varied from 5 to 25 kPa. The authors found that higher temperatures and lower pressures led to higher drying rates. In addition, both blanching and freezing reduced the drying time, being freezing more effective than blanching in this sense. The experimental data were well modeled by two fashions of the Fick's second law of diffusion, viz. considering the occurrence of shrinkage or not. The latter is presented below for unidirectional moisture diffusion through one side of a slab [7]:

$$MR = \frac{8}{\pi^2} \sum_{n=0}^{\infty} \frac{1}{(2n+1)^2} \exp\left(-[2n+1]^2 \pi^2 \frac{D_{eff} \cdot t}{L^2}\right) \tag{1}$$

where M_R is the dimensionless moisture ratio; D_{eff} is the effective moisture diffusivity $(m^2 \, s^{-1})$, t is time (h), L is the thickness of the slab (m) and n is a positive integer. The dimensionless moisture ratio can be defined as

$$MR = \frac{M - M_e}{M_0 - M_e} \qquad (2)$$

where M is the moisture content of the product (kg water kg dry solid^{-1}) at any time; M_e is the moisture content of the product when equilibrium is attained; and M_0 is the initial moisture content of the product. The obtained effective moisture diffusivities ranged from 2.01×10^{-9} to $5.70 \times 10^{-9} \, m^2 \, s^{-1}$ when shrinkage was neglected and from 1.13×10^{-9} to $3.90 \times 10^{-9} \, m^2 \, s^{-1}$ when shrinkage was taken into account. Summarizing, the use of freezing pretreatment followed by drying under high temperature and low pressure leads to high drying rates. In addition, when product shrinkage is taken into account, the values of effective diffusivity are lower.

Arévalo-Pinedo and Murr [3] studied the vacuum drying kinetics of pre-treated and untreated carrot and pumpkin slabs. The pre-treatments consisted in blanching in hot water at 95 °C for 5 min or freezing at −20 °C for 3 h. The drying temperature varied from 50 to 70 °C and the absolute pressure in the drying chamber varied from 5 to 25 kPa. The drying curves presented a long falling rate period. The faster drying process was the one where freezing was used as pre-treatment, followed by the one where blanching was used and then the control (no pre-treatment). This behavior was attributed to the cell disruption that takes place when the food is frozen, allowing moisture to be removed more easily during the subsequent drying. The drying kinetics was adequately modeled by the Fick's second law of diffusion for slabs (Eq. 1), with high values of R^2 and low values of error of prediction. Pumpkin effective moisture diffusivities ranged from 2.01×10^{-9} to $5.70 \times 10^{-9} \, m^2 \, s^{-1}$, while carrot effective moisture diffusivities ranged from 1.273×10^{-9} to $4.844 \times 10^{-9} \, m^2 \, s^{-1}$. Such variations are related to various drying temperatures and pressures. In sum, freezing proved better than blanching for enhancing the rates of carrot and pumpkin vacuum drying and the observed effective moisture diffusivities were higher for pumpkin in detriment of carrot.

Jena and Das [11] studied the vacuum drying of coconut presscake, a by-product of the coconut processing industry. The main contribution of this study was the development of a new thin-layer drying model, which adequately fitted the changes in coconut presscake moisture content during vacuum drying. The new model, called "Jena and Das", is

$$MR = a\exp\left(-k\theta + b\sqrt{\theta}\right) + c \qquad (3)$$

where M_R is the dimensionless moisture ratio presented in Eq. 2; a, b and c are model constants; and θ is time (s). Relative deviations lower than 15 % were obtained when using the new model. Experiments were carried out by using a 3^2 factorial design where sample thickness (2–4 mm) and drying temperature (65–75 °C) were varied. Pressure was fixed at 62 ± 3 mmHg. Faster drying was associated with high temperature and low thickness. In addition, values of effective moisture diffusivity were calculated by using Fick's second law for slabs (Eq. 1). Effective diffusivities ranged from 7.026×10^{-10} to 3.326×10^{-9} m^2 s^{-1} for various temperatures and thicknesses. Concluding, the Jena and Das model showed suitable for fitting the drying kinetics of coconut presscake and the use of high sample thickness and high drying temperature enhanced the effective moisture diffusivities.

Wu et al. [30] studied the vacuum drying characteristics of eggplant slices (*Solanum melongena*), an important food in several countries. Such as other vegetables, eggplants present short shelf life due to their high water activity. In that study, the effect of variable drying temperatures (30–50 °C) and pressures (2.5–10 kPa) on the drying kinetics was evaluated. The drying kinetics was tentatively fitted by three well-known mathematical models and by a new model that was proposed. The proposed model is shown below:

$$MR = at^2 + bt + 1 \tag{4}$$

where MR is the dimensionless moisture ratio defined as in Eq. 2; a and b are model constants and t is time (h). The effect of moisture on shrinkage was modeled too. Effective moisture diffusivity was evaluated by using Eq. 1, and their dependence on temperature was estimated on the basis of the following Arrhenius-type equation:

$$D_{eff} = D_0 \exp\left(-\frac{E_a}{RT}\right) \tag{5}$$

where D_{eff} is the effective moisture diffusivity (m^2 s^{-1}); D_0 is the pre-exponential factor (m^2 s^{-1}); E_a is referred to as activation energy for moisture diffusion (kJ mol^{-1}); R is the universal gas constant (8.315×10^{-3} kJ mol^{-1} K^{-1}); and T is the absolute temperature (K). The obtained data showed that an increase in drying temperature resulted in a lower drying time. On the other hand, drying chamber pressure did not affect the drying time. Among the drying models used for fitting the drying curve, the proposed polynomial model provided the best result. The effect of moisture content on the shrinkage was well described ($R^2 > 0.98$) by linear models. It was observed that the lower the moisture content, the higher the shrinkage. Effective diffusivities of eggplant slices ranged between 1.6×10^{-9} and 3.4×10^{-9} m^2 s^{-1} as function of variable temperature and pressure. The activation energy for moisture diffusion was presented per unity of mass, i.e., 1,640 kJ/kg. Considering the molecular weight of water as 18 g mol^{-1}, the activation energy was 29.52 kJ mol^{-1}.

Sahari et al. [27] optimized the vacuum drying of dates in order to obtain high quality date powder. Dates are important to the economy of Middle East countries and this was claimed to be the first time that the vacuum drying of dates was studied. The drying temperature varied from 85 to 100 °C and the vacuum pressure varied from 15.3 to 60.1 cmHg. In addition, five different thicknesses of product were tested: 0.2, 1.0, 2.0, 2.5 and 3.0 cm. As a result, a temperature of 85 °C, a pressure of 54.6 cmHg and a product thickness of 1 cm were selected as the optimum conditions. When using the optimum drying conditions, a final product with good smell and color was obtained and the process proved to be cost-effective.

Amellal and Benamara [1] studied the vacuum drying of date pulp cubes from three different varieties, emphasizing the effect of different drying temperatures (60–100 °C) on the drying kinetics of the product. Pressure was fixed at 200 mbar. The Newton [16] and the Henderson-Pabis [8] drying models were used to fit the data. They are respectively presented below:

$$MR = \exp(-kt) \tag{6}$$

$$MR = a\exp(-kt) \tag{7}$$

where MR is the dimensionless moisture ratio presented in Eq. 2; k is the drying rate constant (min^{-1}); a is a model constant and t is time (min). In addition, effective diffusion coefficients were obtained by using Eq. 1 and their temperature dependency was estimated by means of Eq. 5. As expected, an increase in temperature from caused an increase in drying rates. The Henderson-Pabis model was superior to the Newton model as for fitting the kinetics, which was denoted by higher R^2 value and lower mean relative error. The effective diffusion coefficients varied between 2.72×10^{-11} and 1.0×10^{-10} m^2 s^{-1}, being this variation with respect to different date varieties and drying temperatures. The activation energies varied between 13.29 and 24.70 kJ mol^{-1} for different varieties of date. In conclusion, vacuum drying proved a suitable process for increasing dates shelf life and the drying kinetics was well described by the Henderson-Pabis model.

Lee and Kim [13] carried out a comprehensive study on the vacuum drying of Asian radish slices, a popular vegetable in South Korea. The effect of different drying temperatures (40–60 °C) and sample thicknesses (4–6 mm) on drying kinetics/drying rate was compared. Pressure was fixed at 0.1 mPa. Nine thin-layer drying models were tentatively used to fit the changes in radish moisture during drying. The effective diffusivities for various drying conditions were estimated by using Fick's second law for slabs (Eq. 1) and their dependence on temperature was estimated by using an Arrhenius-type relationship (Eq. 5). Those authors observed that working with high temperature and low slice thickness yielded high drying rates. The logarithmic model [31] was found to best represent the drying kinetics, yielding high values of coefficient of determination and low values of root mean square error, mean relative percent deviation and reduced chi-square. Such model is presented below:

$$MR = a\exp(-kt) + c \tag{8}$$

where MR is the dimensionless moisture ratio presented in Eq. 2; k is the drying rate constant (min^{-1}); a and c are model constants and t is time (min). The values of effective diffusivity ranged from 6.92×10^{-9} to 14.59×10^{-9} m^2 s^{-1}, being this variation with regard to various product thicknesses and drying temperatures. Increase in temperature promoted an augment in the values of effective diffusivity, while increase in thickness promoted a decrease in these values. A linear regression was successfully applied to relate effective diffusivity and temperature for the two thicknesses tested. In this way, it became possible to estimate the effective diffusivity of Asian radish slices on the basis of drying temperature for process designing purposes. The values of activation energy ranged from 16.49 to 20.26 kJ mol^{-1}, for different thicknesses. Summarizing, the logarithmic model provided the best fit for the vacuum drying kinetics of Asian radish slices and the effective diffusivities and activation energies were within the range usually observed for foods.

Arévalo-Pinedo et al. [4] compared the effect of variable vacuum drying conditions on the values of effective diffusivity of carrot as obtained by using Fick's diffusional model with and without shrinkage. In the lab, the carrots were sliced and then, pretreated by either blanching or freezing, in an attempt to improve the drying rate. The vacuum drying process was carried out at a temperature of 50, 60 or 70 °C and a pressure of 5, 15 or 25 kPa. For modeling the drying kinetics without considering product shrinkage, unidirectional moisture diffusion in a flat plate (slab) was considered (Eq. 1), as based on Fick's second law of diffusion. For considering shrinkage, the above mentioned diffusional model was changed by adding the density of the dry solid to it (see [23] for more details). Results showed that both pretreatments increased the drying rate, even though frozen/thawed carrots were dried faster than blanched/cooled carrots. Such behavior was attributed to easier moisture removal due to cell disruption caused by freezing. Higher temperatures and lower pressures led to higher drying rates, which were associated with higher values of effective moisture diffusivity. This behavior was attributed to the puffed structure formed in the food during low pressure drying. Considering the different temperatures, pressures and pretreatments, the values of effective diffusivity ranged from 1.27×10^{-9} to 4.84×10^{-9} m^2 s^{-1} when shrinkage was neglected and from 1.11×10^{-9} to 3.40×10^{-9} m^2 s^{-1} when shrinkage was considered. To sum up, it can be affirmed that combining freezing pretreatment, high drying temperatures and low drying pressures leads to high values of effective moisture diffusivity and high drying rates during the vacuum drying of carrots.

Artnaseaw et al. [5] studied the heat pump vacuum drying of mushrooms and chilies from Thailand. As mentioned by the authors, in a vacuum drying process heat can be obtained by several ways, such as electric sources, microwaves and heat pumps. The latter was chosen for that study. For this purpose, an experimental apparatus was constructed. Different temperatures (55–65 °C) and pressures (0.1–0.4 bar) were used. In most of the cases, the use of higher temperatures and lower pressures yielded increased drying rates. Among eleven thin layer drying models used to fit the drying kinetics, the Midilli model [18] presented the best results, i.e., high values of R^2, low values of reduced chi-square and low values of root mean square error. The Midilli model is presented below:

$$M_R = a\exp\left[-k\left(t^n\right)\right] + bt \tag{9}$$

where M_R is the dimensionless moisture ratio presented in Eq. 2; a, k, n and b are model constants and t is time (h). The four constants of the Midilli model were obtained by using regression analysis and expressed as a function of drying temperature or drying pressure. For this purpose, polynomial models provided a good fit (R$^2 > 0.98$). Some quality parameters, viz. color and rehydration capacity were measured in the dried product. In this sense, higher drying temperatures led to lower color degradation, which was attributed to shorter drying times. In addition, lower drying pressures led to lower color degradation too, which was attributed to low oxygen concentration in the drying chamber. In time: oxygen participates in the enzymatic browning reaction. The rehydration capacity decreased with an augment in drying pressure, but no effect of temperature was observed. In sum, this study established process conditions suitable for producing dried mushrooms and chilies by vacuum drying with a proper quality.

Lee and Kim [14] performed the dehydration under controlled conditions of water dropwort (*Oenanthe javanica* DC.), an herb that is used in Korean soups and stews. The work aimed at evaluating the effect of various drying temperatures and pretreatments on drying kinetics and product color parameters. The drying temperature varied between 50 and 70 °C. The pretreatments consisted in blanching at 80 °C during 2 min or dipping in a 1 % potassium meta bisulphate (KMS) aqueous solution for 3 min. Pressure was fixed at 0.1 mPa. Results showed that the use of higher temperatures led to higher drying rates, as expected. Blanched samples were dried faster than control while 1 % KMS treated samples were dried slower than control. The drying kinetics was tentatively fitted by nine thin layer drying models. The goodness of the fit was evaluated by taking into account the values of coefficient of determination, reduced chi-square and root mean square error. The Page model [22] proved to be the most appropriate for explaining the moisture loss of water dropwort during vacuum drying. The Page model can be expressed as:

$$MR = \exp\left(-kt^n\right) \tag{10}$$

where MR is the dimensionless moisture ratio presented in Eq. 2; k and n are model constants and t is time (min). The drying process was shown to occur in the falling rate period, i.e., no constant-rate drying period was observed. Regarding color, the herbs turned darker and lost some of their green-yellow color, i.e., decrease in L* and b* values and increase in a* value. The total color change was more pronounced for 1 % KMS treated samples. Summarizing, the authors recommended that the vacuum dried water dropwort be obtained by blanching followed by drying at 60 °C, which yields a final product with the lowest color degradation.

Mitra et al. [19] optimized the vacuum drying of onion slices by using the response surface methodology. This was a significant study given the wide culinary use of the selected raw material. A factorial design was used to investigate the effect of drying temperature (50–70 °C), slice thickness (1–3 mm) and presence/absence of pretreatment (5 % NaCl plus 0.2 % $K_2S_2O_5$ for 15 min) on quality features of the final product, namely moisture content, color, flavor and rehydration capacity. Second and first order polynomial models successfully explained the effect of the factors on the responses, as denoted by high values of coefficient of determination and low values of coefficient of variation. Optimum quality, expressed as low moisture content, low color change, high flavor retention and high rehydration capacity, was obtained by using the following process conditions: pretreatment followed by slicing the onions to 4.95 mm thickness and drying them at 58.66 °C and 50 mmHg of absolute pressure.

Mitra et al. [20] studied the vacuum drying of onion emphasizing the effect of different process temperatures, slice thicknesses and the use of a pretreatment on the drying kinetics. Thin layer drying models were tentatively used to fit the drying data. Furthermore, the effective diffusivity under different conditions was calculated (Eq. 1) and its dependence on temperature and thickness was investigated. Those authors observed that, at a fixed pressure (50 mmHg), a rise in drying temperature from 50 to 70 °C caused an increase in drying rate. On the other hand, the pretreatment tested (5 % NaCl and 0.2 % $K_2S_2O_5$) did not affect the drying kinetics of thin slices (1 and 3 mm), only of thick slices (5 mm). Comparing four drying models, the Page model (Eq. 10) was found to be the most proper to fit the drying kinetics, yielding high values of R^2 and low values of root mean square error and reduced chi-square. The effective diffusivity as calculated by Fick's second law (Eq. 1) ranged from 1.32×10^{-10} to 1.09×10^{-9} m^2 s^{-1}, values that are related to different thicknesses and temperatures. The dependence of effective diffusivity on temperature and thickness was more properly described by an Arrhenius type equation when compared to a regression model. To sum up, the Page model showed to be proper for describing the moisture variation of onion slices during vacuum drying and the effective diffusivities obtained were within the general range for foods.

Ashraf et al. [6] modeled the vacuum drying of date paste and investigated the effect of variable drying conditions on drying rates. The variables tested were drying temperature (60–80 °C) and sample thickness (1–2 cm). The absolute pressure was kept at 20 kPa. In addition, effective diffusivities were calculated (Eq. 1) and their dependence on temperature was estimated (Eq. 5). Results showed that the Jena and Das, the Verma and the modified Henderson-Pabis models were the most suitable for estimating the changes in moisture content of date paste, yielding high values of coefficient of determination and low values of reduced chi-square and root mean square error. Since the Jena and Das model has been presented before (Eq. 3), only the Verma [29] and the modified Henderson-Pabis [11] models will be shown below, respectively:

$$MR = a\exp(-kt) + (1 - a)\exp(-gt) \tag{11}$$

$$MR = a\exp(-kt) + b\exp(-gt) + c\exp(-ht) \tag{12}$$

where MR is the dimensionless moisture ratio presented in Eq. 2; a, b, c, g, h and k are model constants and t is time (min). It was observed that the use of low product thickness and high drying temperatures resulted in high drying rates. The effective diffusivities as calculated by the Fick's second law of diffusion for a slab (Eq. 1) presented values between 6.085×10^{-8} and 4.868×10^{-7} m^2 s^{-1}. When high temperatures and high sample thickness were used, high values of effective diffusivities were obtained. In time: thin slabs are more subjected to surface hardening, which impairs the moisture diffusion. Finally, the temperature dependence of the effective diffusion coefficient was well described by an Arrhenius type equation (Eq. 5). The activation energy ranged from 33.71 to 54.96 kJ mol^{-1} for variable product thickness. In conclusion, three models presented a proper fit for the kinetic data and the values of effective diffusivity and activation energy were slightly above the usual values for foods.

Lee et al. [15] studied the vacuum drying of *Salicornia herbacea* L., a salt-tolerant herb which presents functional properties. The drying kinetics was obtained for 50, 60, 70 and 80 °C. Absolute pressure was kept at 0.1 mPa. The drying rates were calculated. Thin layer drying models were used to fit the data. The dependence of the drying constant of the best model on temperature was estimated by an Arrhenius type equation. With regard to quality evaluation, the color of the dried/powdered herbs and the image of the dried herbs were ana-lyzed. It was observed that the higher the drying temperatures, the higher the dry-ing rates. Drying rates decreased with the advance of the drying process. In time: the residual water present in the food at the end of drying is strongly bounded to the other food components, taking longer to be removed. Among seven mathemat-ical models used, the logarithmic model (Eq. 8) was found to be the most suitable for fitting the drying kinetics. This finding was supported by the values of coeffi-cient of determination, root mean square error, mean relative percent deviation and reduced chi-square. The drying constant "k" in the logarithmic model was shown to be temperature dependent, with an activation energy of 15.02 kJ mol^{-1}. The dried product, which was originally dark green, became light green after drying, as expressed by increase in lightness (L*), yellowness (b*) and saturation (C*) and decrease in redness (a*) and hue angle (h*). The total color difference (ΔE) between the fresh and the dried sample varied between 26.40 and 27.03 for the various temperatures tested. The images obtained with a scanning electron micro-scope presented a high amount of wrinkles, especially for higher drying tempera-tures. On the other hand, pores were not found. Taking into consideration proper color preservation and short drying time, the process conducted at 80 °C was found to be the most suitable.

Reis et al. [24] elucidated and modeled the vacuum drying kinetics of yacon (*Smallanthus sonchifolius*) and studied the effect of variable drying conditions on the final product fractal dimension and rehydration ratio. The tubers were sliced at thicknesses ranging from 0.2 to 0.6 cm and pretreated by immersion in citric acid solutions (0.2–1.0 % w/w). Drying temperature varied from 45 to 65 °C. Pressure was fixed at 7.6 kPa. Among the models tested, the modified Page model [21] was found to provide the best fit to the experimental data. The modified Page model can be represented as

$$M_R = \exp\left[(-kt)^n\right] \tag{13}$$

where M_R is the dimensionless moisture ratio presented in Eq. 2; k and n are model constants and t is time (h). Among the variable drying conditions, it was observed that the higher the drying temperature, the lower the product frac-tal dimension. The use of the fractal dimension to express the quality of foods is recent. This novel quality parameter presents the advantages of being easily meas-ured by an image analysis and providing a comprehensive profile of one food appearance, which is not achievable with a colorimeter. Increase in slice thickness was found to negatively affect the rehydration capacity of the yacon slices, which was justified by the difficulty of the rehydration water to reach the inner portions of the thicker slices. In addition, significant correlations were obtained between fractal dimension and color, moisture content and rehydration capacity, suggesting that these typical quality features could be substituted by an image analysis.

Reis et al. [25] studied the changes in color and texture of yacon slices during vacuum drying and the effect of different drying conditions on such quality features. In addition, the vacuum drying conditions were optimized toward color quality by means of the response surface methodology. The tubers were sliced to 0.2, 0.4 and 0.6 cm and pretreated by immersion in citric acid solutions (0.2–1.0 % w/w). Drying temperature was set at 45, 55 or 65 °C. Pressure was fixed at 7.6 kPa. Yacon color was expressed by using CIE L*a*b* and CIE L*C*h* color spaces. The authors observed that the product color changed during drying as per a decrease in lightness (L*) and an increase in redness (a*) and yellowness (b*). The product hardness did not present significant changes during initial stages of drying, followed by softening and hardening. Such results are attributed a loss of the original vegetable cells turgor followed by formation of external crust at the end of drying. Results showed that the use of high drying temperatures (58–65 °C), high citric acid concentrations (0.8–1.0 g/100 g) in the pretreatment solution and low thicknesses (≤0.4 cm) yielded a product of ideal color, which was represented by high lightness (L*), high yellowness (b*) and high colorfulness (C*).

Thorat et al. [28] studied the vacuum drying of ginger (*Zingiber officinale* R.). The influence of different vacuum drying temperatures (40–65 °C) on the drying kinetics of ginger slices was elucidated. The absolute drying pressure was 8 kPa. The drying curve was modeled by using five thin layer drying models. The effective diffusion coefficient was estimated by using Eq. 1 and the effect of temperature on it was tentatively described by an Arrhenius-type relationship (Eq. 5). It was observed that the use of higher temperatures (65 °C) led to higher drying rates compared to the use of lower temperatures (40 °C). The two-term mathematical model provided the best fit to the drying curve, as confirmed by high values of R^2 and low values of reduced chi-square and root mean square error. The two-term model (Henderson 10) is represented as follows:

$$MR = a\exp(-kt) + b\exp(-k_0 t) \tag{14}$$

where MR is the dimensionless moisture ratio presented in Eq. 2; a, b, k and k_0 are model constants and t is time (min). The drying curve could be divided in two falling rate periods, while no constant rate period was observed. The effective diffusion coefficient varied roughly from 1.9 to 4.8×10^{-8} m^2 s^{-1} and proved to be temperature-dependent. The calculated activation energy for water diffusion was 35.7 kJ mol^{-1}. Concluding, the vacuum drying curve of ginger slices was adequately fitted by the two-term model and the values of effective moisture diffusivity and activation energy obtained for the process were within the typical range for food products.

The vacuum drying of loquat fruit (*Eriobotrya japonica* Lindl.) was studied by Saberian et al. [26]. More specifically, the influence of different drying temperatures on the drying time, drying rate and effective moisture diffusivity was evaluated and the drying was tentatively modeled by nine thin layer drying models. Vacuum drying was performed at 60, 70 and 80 °C and a vacuum of 52 cmHg (absolute pressure of ~32 kPa). Results showed that the higher the temperature, the higher the drying rates and the shorter the drying time, as expected. Effective moisture diffusivity as calculated by using Eq. 1 ranged from 6.87×10^{-10} to 1.29×10^{-9} m^2 s^{-1}, increasing with an increase in temperature. With regard to the loquat fruit drying curves modelling, the Page model (Eq. 10) was the most suitable model, with the highest value of coefficient of determination and the lowest values of root mean square error. In addition, the approximation of diffusion model presented a good fit as well. In conclusion, the vacuum drying of loquat fruit was accelerated by an increase in temperature and could be well described by the Page model.

References

1. Amellal H, Benamara S (2008) Vacuum drying of common date pulp cubes. Dry Technol 26:378–382
2. Arévalo-Pinedo A, Murr FEX (2006) Kinetics of vacuum drying of pumpkin (*Curcubita maxima*): modeling with shrinkage. J Food Eng 76:562–567
3. Arévalo-Pinedo A, Murr FEX (2007) Influence of pre-treatments on the drying kinetics during vacuum drying of carrot and pumpkin. J Food Eng 80:152–156
4. Arévalo-Pinedo A, Murr FEX, Arévalo ZDS et al (2010) Modeling with shrinkage during the vacuum drying of carrot (*Daucus carota*). J Food Process Pres 34:611–621
5. Artnaseaw A, Theerakulpisut S, Benjapiyaporn C (2010) Drying characteristic of Shiitake mushroom and Jinda chili during vacuum heat pump drying. Food Bioprod Process 88:105–114
6. Ashraf Z, Hamidi-Esfahani Z, Sahari MA (2012) Evaluation and characterization of vacuum drying of date paste. J Agric Sci Technol 14:565–575
7. Crank J (1975) The mathematics of diffusion. Claredon Press, Oxford
8. Henderson SM, Pabis S (1961) Grain drying theory I: temperature effect on drying coefficient. J Agric Res 7:85–89
9. Henderson SM (1974) Progress in developing the thin layer drying equation. T ASAE 17:1167–1172
10. Jaya S, Das H (2003) A vacuum drying model for mango pulp. Dry Technol 21:1215–1234
11. Jena S, Das H (2007) Modelling for vacuum drying characteristics of coconut presscake. J Food Eng 70:92–99
12. Karathanos VT (1999) Determination of water content of dried fruits by drying kinetics. J Food Eng 39:337–344
13. Lee JH, Kim HJ (2009) Vacuum drying kinetics of Asian white radish (*Raphanus sativus* L.) slices. LWT-Food Sci Technol 42:180–186
14. Lee JH, Kim HR (2010) Influence of pretreatments on the dehydration characteristics during vacuum drying of water dropwort (*Oenanthe javanica* DC.). J Food Process Pres 34:397–413
15. Lee JH, Kim HJ, Rhim JW (2012) Vacuum drying characteristics of *Salicornia herbacea* L. J Agric Sci Technol 14:587–598
16. Lewis WK (1921) The rate of drying of solid materials. Ind Eng Chem 13:427–443
17. Madamba PS, Liboon FA (2001) Optimization of the vacuum dehydration of celery (*Apium graveolens*) using the response surface methodology. Dry Technol 19:611–626
18. Midilli A, Kucuk H, Yapar Z (2002) A new model for single-layer drying. Dry Technol 20:1503–1513
19. Mitra J, Shrivastava SL, Srinivasa Rao P (2011) Process optimisation of vacuum drying of onion slices. Czech J Food Sci 29:586–594
20. Mitra J, Shrivastava SL, Srinivasa Rao P (2011) Vacuum dehydration kinetics of onion slices. Food Bioprod Process 89:1–9

21. Overhults DD, White GM, Hamilton ME et al (1973) Drying soybeans with heated air. T ASAE 16:195–200
22. Page G (1949) Factors influencing the maximum rates of air-drying shelled corn in thin layer. Dissertation, Purdue University
23. Park KJ (1998) Diffusional model with and without shrinkage during salted fish muscle drying. Dry Technol 16:889–905
24. Reis FR, Lenzi MK, Muñiz GIB et al (2012) Vacuum drying kinetics of yacon (*Smallanthus sonchifolius*) and the effect of process conditions on fractal dimension and rehydration capacity. Dry Technol 30:13–19
25. Reis FR, Lenzi MK, Masson ML (2012) Effect of vacuum drying conditions on the quality of yacon (*Smallanthus sonchifolius*) slices: process optimization toward color quality. J Food Process Pres 36:67–73
26. Saberian H, Amooi M, Hamidi-Esfahani Z (2014) Modeling of vacuum drying of loquat fruit. Nutr Food Sci 44:24–31
27. Sahari MA, Hamidi-Esfahani Z, Samadlui H (2008) Optimization of vacuum drying characteristics of date powder. Dry Technol 26:793–797
28. Thorat ID, Mohapatra D, Sutar RF et al (2012) Mathematical modeling and experimental study on thin-layer vacuum drying of ginger (*Zingiber officinale* R.) slices. Food Bioprocess Tech 5:1379–1383
29. Verma LR, Bucklin RA, Endan JB et al (1985) Effects of drying air parameters on rice drying models. T ASAE 28:296–301
30. Wu L, Orikasa T, Ogawa Y et al (2007) Vacuum drying characteristics of eggplants. J Food Eng 83:422–429
31. Yagcioglu A, Degirmencioglu A, Cagatay F (1999) Drying characteristic of laurel leaves under different conditions. In: Bascetincelik A (ed) Proceeding of the 7th international congress on agricultural mechanization and energy, Adana
32. Zhang M, Li C, Ding X et al (2003) Optimization for preservation of Selenium in sweet pepper under low-vacuum dehydration. Dry Technol 21:569–579

Studies on Freeze-Drying of Foods

Maria Lucia Masson

Abstract Quality and costs are the two words that have risen from any discussion about freeze-drying processes through the years. Optimizing this process always depended mainly on practical experiments and measurement of changes in color, texture, rehydration ratio etc. This chapter has the goal to show the evolution of the studies on freeze-drying of foods. Nowadays, researches are proposing a substitution of traditional methodologies by studies dealing with the influence of glass transition temperature as a parameter to express the quality changes of freeze-dried foods. In addition, many models based in mass and energy balances, including heat transfer, have been developed and are under investigation in order to provide an efficient tool to allow the control and optimization of freeze-drying processes. New technological solutions were studied during the last years, such as the microwave-assisted freeze-drying of foods. By reading the following chapter, it is possible to observe that there has been a great scientific and technical evolution in the study of freeze-drying of foods.

Keywords Freeze-drying · Glass transition temperature · Energy balance · Mass balance

Food drying processes are complex and involve many critical variables for determining the final product quality and shelf-life. Drying temperature and drying time are dependent on each other and are the most important features to determine a good process. So, drying processes conducted at lower temperatures are a goal for food drying, thus avoiding damages associated with product exposition to high temperatures or to moderate temperatures for long times (changes in product physical, chemical and biological properties). The most critical product physical properties are color and structure. Remarkable chemical and biological changes in

M.L. Masson (✉)
Chemical Engineering Department, Federal University of Paraná, Curitiba, Brazil
e-mail: masson@ufpr.br

© The Author(s) 2014
F. Richter Reis (ed.), *Vacuum Drying for Extending Food Shelf-Life*, SpringerBriefs in Applied Sciences and Technology, DOI 10.1007/978-3-319-08207-3_3

Fig. 1 Energy cost
breakdown for freeze-drying
process [13]

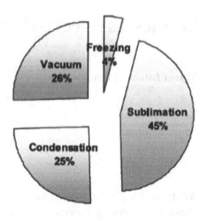

product quality include degradation of sugars, proteins and loss or alteration of aroma compounds. In a way to obtain dry food materials with high quality, many studies were conducted to develop and optimize the process of freeze-drying. Freeze-drying is considered the best method of water removal because, when compared to other methods of food drying, it yields final products of highest quality [6, 11]. The freeze-drying process consists in freezing, vacuum, sublimation and condensing operations, yielding a total energetic consumption share as shown in Fig. 1. This process is based on sublimation for dehydrating a frozen product. The absence of liquid water and the low temperatures required for sublimation are responsible for stopping most of the deteriorative reactions, e.g., microbiological reactions, thus allowing the obtainment of a final product of excellent quality. This is particularly important for value-added foods such as functional foods, baby foods and others special classes of foods, considering that freeze-drying is the most expensive process for manufacturing a dehydrated product [1, 11, 13].

Between 2005 and 2013, researchers of many areas of investigation published 9,362 works (articles, reviews, patents) in the world about freeze-drying, according to a search in Scopus database in year 2014. Among them, 2,836 works were studies dealing with applications of freeze-drying in the area of Food Science and Technology. According to Liu et al. [8], the energy required to remove 1 kg of water by freeze-drying is at least two times higher than that needed in conventional drying operations. Claussen et al. [2] wrote a review reporting values of specific moisture extraction rate in the range of 4.6–1.5 kg of water per kWh for atmospheric freeze-drying with heat pumps and 0.4 or below for industrial vacuum freeze-drying. Some characteristics make freeze drying process expensive, like the fact of being a slow process (generally batch). The process is composed by three stages: freezing, primary drying and secondary drying [11].

Glass transition temperature (T_g) can be defined as the temperature at which an amorphous system changes from the glassy to the rubbery state and is a parameter experimentally determined [13]. This property has been pointed out to be responsible for the deterioration and an indicator of food stability. It has been reported that when temperature of one process exceeds T_g, the quality of the food being processed is seriously altered, as reported by Ratti [13]. A possible relationship between non-enzymatic browning, T_g and temperature is the Williams-Landel-Ferry equation proposed by Williams et al. [20]:

$$\ln \frac{k_{ref}}{k} = \frac{-C_1(T - T_g)}{C_2 + (T - T_g)} \tag{1}$$

where k is the kinetic constant, c_1 and c_2 are empirical constants, T is temperature and T_g is the glass transition temperature.

Santivarangkna et al. [15] in a study about the role of the glassy state on stabilities of freeze-dried probiotics referred glassy state to an amorphous metastable state that resembles a solid but without any long-range lattice order, i.e., a material at the glassy state presents a solid characteristic and appearance but molecular arrangement presented is more typical of liquids. Viscosity at the glassy state is typically $\geq 10^{12}$ Pa s and is temperature-dependent [17]. The most used parameter to describe the glassy state and its transition is the glass transition temperature (T_g), below which materials exhibit extremely high viscosity. Sugars or strong glass-forming polymers have been added in an effort to increase T_g of dried probiotics. Some authors affirmed that a glassy state during the freezing step could inhibit ice crystal formation and a maximally freeze-concentrated solution was formed [14]. This finding suggests that the formation of a maximally freeze-concentrated matrix with entrapped microbial cells is essential for survival of probiotics during freezing [10]. This condition can be obtained at freezing using liquid $N_2(-196 \ °C)$, considering the T_g of sucrose of $-46 \ °C$ and that of the pure water of $-135 \ °C$. Apparently, product temperature during removal of ice crystals by sublimation during freeze-drying should not be higher than T_g and for this reason, solutes with high T_g are added to the food system in order to provide high storage stability. Fonseca et al. [5] proposed to use the collapse temperature, i.e., the maximum temperature preventing the structure of dried product from macroscopic collapse, as the critical temperature for freeze-drying of a cell-containing formulation. This is of economic importance due to the fact that it represents a decrease in primary drying time by about 13 % according to Tang and Pikal [18]. Previous studies showed that the viability of probiotics was influenced by physical state, residence time in rubbery state and moisture content of samples. Stability during storage is critical for probiotics and is not guaranteed by low moisture contents. According to Santivarangkna et al. [15], very little information on sorption isotherm of probiotics was available in comparison to foods and stability is commonly related to survival. Nevertheless, deteriorative reactions, such as lipid oxidation is not stopped by formation of glassy state during storage of dried probiotics.

In studies about freeze drying of foods, Shishehgarha et al. [16] studied the drying kinetics, color, and volume changes of whole and sliced strawberries under various temperatures and found that dehydration time increased proportionally to the thickness of the product, while increase in heating plate temperature markedly reduced it. Freeze drying caused a decrease in hue angle (h*) by 22.5 % at skin and by 42.4 % at pulp, along with a reduction in volume of 8 % in whole and 2 % in sliced strawberries. Marques et al. [9] determined various quality parameters for freeze-dried acerola like deformation by shrinking, water activity as related to different moisture content, glass transition temperature, changes in ascorbic acid content and rehydration ratio. Drying kinetics for crushed and sliced acerola was also determined. They found that sorption isotherms are practically not affected by the freezing technique used, i.e., direct method with liquid nitrogen, direct method with nitrogen gas or placing the samples in a freezer. Freeze-drying tests were performed at a vacuum chamber absolute pressure and a condenser temperature of 1.3×10^{-1} mbar and -30 °C, respectively. The thermocouple probe at the bottom of the tray was used to control and monitor the product temperature. The heat was supplied for sublimation by a heating plate under the tray. During the secondary stage of drying, the product reached a final temperature of about 35 °C and freeze-drying time was approximately 12 h. The freezing method was found important to drying rates with acerola frozen using N_2 vapor presenting higher drying rates than those frozen using liquid N_2, due to size and distribution of crystals of ice formed. Glass transition temperature for dried powder was found equal to -32.1 °C at a moisture content of 0.25 kg/kg dry solid and specific heat values ranged from 0.666 to 0.667 J/g K. They found also that freeze-dried acerola presented minimum shrinkage, high rehydration capacity (a 8.1 kg/kg dry solid rehydrated product moisture content), and good preservation of vitamin C, with a maximum content of 153.4 mg/g dry solid.

Freeze-drying of quince (*Cydonia oblonga*) was studied by Adhami et al. [1], where the drying kinetics determined at various operating conditions. In addition, the effects of initial moisture content, heat load power and initiation time of heat application on drying rates and performance of the drier were investigated. The equation selected to describe the relationship between product moisture content and drying time was:

$$\ln\left(\frac{X}{X_0}\right) = a + bt + ct^2 + dt^3 + et^4 \tag{2}$$

where X is the product moisture content at any time t, X_0 is the initial moisture content and a, b, c, d, and e are the fitting parameters. By the derivation of Eq. 2, the experimental drying rate equation is obtained:

$$\frac{d\left(\frac{X}{X_0}\right)}{dt} = \left(b + 2ct + 3dt^2 + 4et^3\right)\exp\left(a + bt + ct^2 + dt^3 + et^4\right) \tag{3}$$

where the equation parameters are similar to those described for Eq. 2. In addition, those authors proposed models for the drying kinetics of the primary and secondary drying stages, as follows:

$$\frac{dX}{dt} = \left(a + bq^n\right)\left(c + dT^m\right)\left(\frac{X}{X_0}\right)^k \tag{4}$$

Fig. 2 Schematic of the vial geometry with the dimensional coordinate system and the dimensionless coordinate system, after the axial immobilisation of the moving interface: z axial coordinate (m); H position of the moving front (m); ξ^I and ξ^{II} non-dimensional axial-coordinate for detailed model; L total thickness of the product (m); l air gap at the vial bottom (m) [19]

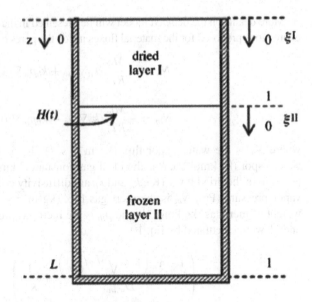

$$\frac{dX}{dt} = a'\left(\frac{X}{X_0}\right)^{n'} - b'\exp\left(\frac{c'}{T} - d'\right)\left(e' + q^{k'}\right) \tag{5}$$

where q is the heat load (W/kg), T is the sample temperature, and a, b, n, c, d, m, k, a′, n′, b′, c′, d′, e′ and k′ are fitting parameters. Equations 2, 3, 4 and 5 were applied to the experimental data obtained for quince, resulting in Eq. 6 for the primary drying stage and Eq. 7 for the secondary drying stage, as follows:

$$\frac{dX}{dt} = \left(0.0797 + 0.0007q^{0.9077}\right)\left(1.1011 + 4.4060T^{-0.3246}\right)\left(\frac{X}{X_0}\right)^{0.7954} \tag{6}$$

$$\frac{dX}{dt} = 1.8224\left(\frac{X}{X_0}\right)^{1.3716} - 29.0836\exp\left(\frac{44.6125}{T} - 8.7070\right)\left(0.5844 + q^{0.6116}\right) \tag{7}$$

where the equation parameters are similar to those described for Eq. 2.

Velardi and Barresi [19] developed and applied mono-dimensional models to the primary drying of the freeze-drying process of bovine serum albumin in vials. Initially, a detailed mono-dimensional model considering mass/energy balances in the dried layer and at the sublimating interface was developed. Based on this first detailed model, two simplified models were developed for allowing their application in on-line monitoring. Those authors justified their study by affirming that multi-dimensional models are quite complex and their solution is highly time consuming and make real-time applications unsuitable. Therefore, radial gradients were neglected. The detailed mono-dimensional model proposed introduced energy balance describing heat transfer in the glass vial (Fig. 2), in presence of radiation from the freeze-dryer chamber.

More details on the detailed model will be presented in the following paragraphs. The expressions proposed for the material fluxes inside the pores of the dried layer were:

$$N_w = -\frac{M_w}{RT_l}(k_a \nabla p_w + k_b p_w \nabla p) \tag{8}$$

$$N_{in} = -\frac{M_{in}}{RT_l}(k_c \nabla p_{in} + k_b p_{in} \nabla p) \tag{9}$$

where N_w is the water vapor flux (kg m^{-2} s^{-1}), M_w is the molecular weight of water vapor (kg kmol^{-1}), R is the ideal gas constant (J kmol^{-1} K^{-1}), T_l is the temperature in the dried layer (K), k_a and k_b are diffusivity coefficients, p_w is the water vapor pressure (Pa), N_{in} is the inert gas flux (kg m^{-2} s^{-1}), M_{in} is the molecular weight of inert gas (kg kmol^{-1}) and p_{in} is the inert gas pressure (Pa). Properties *ka* and *kb* were calculated by Eq. 10:

$$k_a = \left(\frac{1 - y_w\left(1 - \sqrt{M_w/M_{in}}\right)}{D_{w,in}} + \frac{1}{K_w}\right)^{-1},$$

$$k_b = k_a \frac{K_{in}}{D_{w,in}p} + \frac{\kappa}{v}, \quad k_c = k_a\sqrt{M_w/M_{in}}, \tag{10}$$

$$K_w = c_1\sqrt{\frac{RT}{M_w}}, \quad K_{in} = c_1\sqrt{\frac{RT}{M_{in}}}, \quad D_{w,in} = \delta_{w,in}\frac{\varepsilon p}{\tau p}$$

where y_w is the molecular fraction of water vapor, K_w and K_{in} are the Knudsen diffusivity of water vapor and inert gas, respectively (m^2 s^{-1}), $D_{w,in}$ is the effective binary diffusion coefficient (m^2 s^{-1}), v is the dynamic viscosity (kg m^{-1} s^{-1}), κ is a constant dependent only upon structure of porous medium (m^2), c_1 is a constant dependent only upon structure of porous medium and giving relative Knudsen flow permeability (m), $\delta_{w,in}$ is the free binary diffusion coefficient (m^2 s^{-1}), εp is the porosity of the dried layer, τp is the tortuosity factor, and the other parameters have been described before.

The water sorption/desorption rate at the interface between pore surface and vapor phase was described by:

$$\frac{\partial \rho_{sw}}{\partial t} = k_{sw,1}\rho_w\left(\rho_{sw}^* - \rho_{sw}\right) - k_{sw,2}\left(\rho_{sw} - \rho_{sw}^*\right) \tag{11}$$

where ρ_{sw} is the mass concentration of bound water (kg m^{-3}), t is time (s), $k_{sw,1}$ is the kinetic rate constant for adsorption (m^3 kg^{-1} s^{-1}), $k_{sw,2}$ is the kinetic rate constant for desorption (s^{-1}) and ρ_{sw}^* is the equilibrium bound water mass concentration (kg m^{-3}).

An energy balance was performed for the dried and for the frozen layers. Mass and heat balances at the sublimating interface were also performed. The former provided the value of the velocity of the sublimating moving interface (related to the rate of sublimation). The latter gave the enthalpy change across the interface. An energy balance in the vial was performed taking into account terms of heat accumulation and conduction, heat transfer by radiation within the chamber environment and internal heat exchange within the product. A comparison between simulation and experimental results was also carried out. Experimental data were obtained through a series of freeze-drying cycles in a pilot unit capable of providing a $-65\,°C$ condenser temperature. The vials had an internal radius equal to 7.125×10^{-3} m, mean thickness equal to 1.0×10^{-3} m and total volume of 4 ml. The material employed was a 5 % solution of bovine serum albumin (BSA), buffered with tris-HCl 0.1 M. Simulations showed that radial effects could be neglected, and the effect of heat conduction in the vial sidewall is as significant as the contribution of the radiation from the chamber to the drying time and maximal interface temperature.

Two simplified models were developed on the basis of the detailed model: the first comprised an energy balance for the frozen product and a mass balance for the water vapor inside the product. Pseudo-stationary conditions were assumed. The second simplified model comprised heat transfer through the dried and through the frozen layer and along the vial glass as well, while mass transfer modelling was similar to that used for the first simplified model. Results showed that the simplified models allowed the explanation of the behavior of BSA solutions during freeze-drying with relative simplicity. This finding led to the conclusion that the detailed model was useful for serving as a basis for the simplified models, which could be used, for example, for designing soft-sensors and algorithms for freeze-drying monitoring and control.

Energy consumption is always one important topic under investigation. Fissori et al. [4] focused on this aspect. According to those authors, freeze-drying is initially conducted at temperatures as low as $-50\,°C$, where mainly free water is frozen. Subsequently the pressure is lowered to provide ice sublimation, comprising the primary stage of the process. Afterwards, temperature is increased at the freeze-drying chamber and pressure can be reduced to allow enhanced desorption of bound water. Studies dealing with freeze-drying costs have been conducted for a long time, in an attempt to turn this process more cost-effective. More specifically, those authors studied the freeze-drying of coffee extract in trays, determining model parameters from experimental investigation and identifying the best operational conditions by using design space calculation and exergetic analysis. The simplest of the one-dimensional models proposed by Velardi and Barresi [19] was used to describe the dynamic of process. As mentioned before, such model is composed by energy balance for the frozen product and a mass balance for the water vapor inside the product. Energy accumulation, mass accumulation, presence of an inert gas and heat transfer of the side wall of the container were neglected. The apparatus used in this study consisted of two units in series, drying chamber and condenser chamber, connected by a short duct. The experiment was performed using commercially available freeze-dried coffee and de-ionized water as to form a solution containing 25 % of solute and whose glass transition temperature was measured using a differential scanning calorimeter (DSC). According the obtained results, high shelf temperature $(-5\,°C)$ and low chamber pressure (5 Pa) led to reduced drying time. In terms of exergy, a property that reflects one process energy efficiency, the results were the opposite, i.e., low shelf temperature (-20 °C) and high chamber pressure (30 Pa) led to reduced exergy losses. Those results were affected by thermal properties of the system, values for heat and mass transfer coefficients and showed valid only for freeze-drying of coffee.

Condensers play an important role in freeze drying: they have to be very efficient because the maintenance of pressure in the chamber at desired value depends on them. Thus, factors that affect the condenser efficiency such as flow dynamics of sublimated vapour, chamber design, duct size, location and type of closing valve are of remarkable importance. Petitti et al. [11] investigated the effect of flow dynamics and ice deposition on the efficiency of condensers used in freeze-drying equipment by employing mathematical modelling and Computational Fluid Dynamics (CFD) simulation. They examined the following equipment: a laboratory-scale freeze-dryer and an industrial-scale condenser. Assuming that deposition process could be modelled as a first-order process, a parametric study on kinetic constant was carried out. Results showed that the proposed mathematical models provided a good fit from a qualitative viewpoint. As regards condenser efficiency, the use of low inert gas (nitrogen) mass fraction in the inlet (1 %) and low sublimation flow rate led to increased efficiency.

Freeze-dried products have high porosity, which expose surface of food product to O_2 action. This action is particularly important when lipids are contained in the food. Rahman et al. [12] studied the effects of moisture content and storage temperature on the lipid oxidation of freeze dried grouper fish. They observed three trends in the changes of peroxide value depending on the moisture content and storage temperature. They employed two models to fit experimental data of sorption isotherms: BET model and GAB model (Eq. 1 in chapter Comparative Studies). The former is presented below:

$$M_w = \frac{M_b C_b a_w}{(1 - a_w)[1 + (C_b - 1)a_w]} \qquad (12)$$

where M_w is the product moisture content (g water/g dry solids), M_b is the BET monolayer moisture content (g water/g dry solids), C_b is a constant related to the net heat of sorption and a_w is the product water activity.

Glass transition was measured by DSC measurements, and a reaction kinetics model was used to describe parallel reactions occurring in food matrix, considering:

$$A \xrightarrow{k_1} B \xrightarrow{k_2} C$$

where A, B and C are respectively fat, peroxide and degradation products from peroxide; and k_1 and k_2 are reaction rate constants (day^{-1}). According to Giannakourou and Taoukis [7], the analytical solution for the differential rate equation when external conditions are constant is:

$$[B] = [B]_0 \exp(-k_2 t) + k_1 [A]_0 \left[\frac{\{\exp(-k_1 t)\}\{\exp(-k_2 t)\}}{k_2 - k_1} \right] \qquad (13)$$

where the peroxide value is B, which varies with time t.

Fig. 3 Rate constant (k_2) as function of T/T_g for sample containing 5 and 20 g/100 g sample [12]

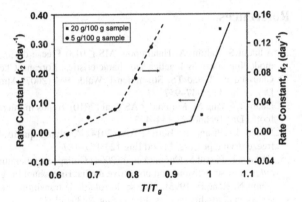

Fig. 4 Rate constant (k_2) as function of X_w/X_b for different temperatures of storage (40, 25, 5, −20 and −40 °C) [12]

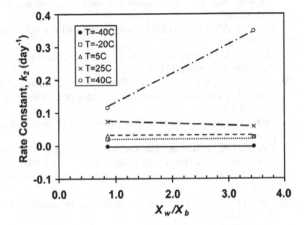

They concluded, from sorption/desorption isotherms, values of peroxide and glass transition measured for grouper fish, that the water activity as expressed by X_w/X_b, and the glass transition temperature as expressed by T/T_g, were important to determine the stability of freeze-dried fish muscle to fat oxidation. Plots of reaction rate constant k_2 as a function of T/T_g (Fig. 3) and X_w/X_b (Fig. 4) were used by authors to confirm the above-mentioned results.

Freeze-drying was commonly accepted as a slow process until microwave-assisted freeze drying (MFD) was developed. MFD is a rapid dehydration technique that can be applied to certain foods, particularly to seafood, solid soup, fruits and vegetables and presents potential in the food industry according to Duan et al. [3]. Advantages of this process include improved product quality, energy savings, shorter drying time and flexibility in production. However, its application is limited due to high startup costs and complex technology compared to conventional freeze drying. MFD was always studied in comparison to other drying techniques. Therefore, more details on this technique, including an MFD apparatus will be presented in last chapter that deals with comparative studies.

References

1. Adhami S, Rahimi A, Hatamipour MS (2013) Freeze drying of quince (*Cydonia oblonga*): modelling of drying kinetics and characteristics. Korean J Chem Eng 30:1201–1206
2. Claussen IC, Ustad TS, Strømmen I, Walde PM (2007) Atmospheric freeze drying. A Rev Dry Technol 25:947–957
3. Duan X, Zhang M, Mujumdar AS et al (2010) Trends in microwave-assisted freeze-drying of foods. Dry Technol 28:444–453
4. Fissori D, Pisano R, Barresi AA (2014) Applying quality-by-design to develop a coffee freeze-drying process. J Food Eng 123:179–187
5. Fonseca F, Passot S, Cunin O et al (2004) Collapse temperature of freeze-dried *Lactobacillus bulgaricus* suspensions and protective media. Biotechnol Prog 20:229–238
6. Genin N, René F (1995) Analyse du rôle de la transition vitreuse dans les procédés de conservation agro-alimentaires. J Food Eng 26:391–408
7. Giannakourou MG, Taoukis PS (2006) Reaction kinetics. In: Sablani SS, Datta AK, Rahman MS, Mujumdar AS (eds) Handbook of food and bioprocess modeling techniques. CRC Press, Boca Raton
8. Liu Y, Zhao Y, Feng X (2008) Exergy analysis for a freeze-drying process. Appl Therm Eng 28:675–690
9. Marques LG, Ferreira MC, Freire JT (2007) Freeze-drying of acerola (*Malpighia glabra* L.). Chem Eng Process 46:451–457
10. Pehkonen KS, Roos YH, Miao S (2008) State transitions and physicochemical aspects of cryoprotection and stabilization in freeze-drying of *Lactobacillus rhamnosus* GG (LGG). J Appl Microbiol 104:1732–1743
11. Petitti M, Barresi A, Marchisio DL (2013) CFD modelling of condensers for freeze-drying processes. Sadhana 38:1219–1239
12. Rahman MS, Al-Belushi RM, Guizani N et al (2009) Fat oxidation in freeze-dried grouper during storage at different temperatures and moisture contents. Food Chem 114:1257–1264
13. Ratti C (2001) Hot air and freeze-drying of high-value foods: a review. J Food Eng 49:311–319
14. Roos YH (1997) Frozen state transitions in relation to freeze drying. J Therm Anal 48:535–544
15. Santivarangkna C, Aschenbrenner M, Kulozik U (2011) Role of glassy state on stabilities of freeze-dried probiotics. J Food Sci 76:R152–R156
16. Shishehgarha F, Makhlouf J, Ratti C (2002) Freeze-drying characteristics of strawberries. Dry Technol 20:131–145
17. Slade L, Levine H (1991) Beyond water activity: recent advances based on an alternative approach to the assessment of food quality and safety. Crit Rev Food Sci 30:115–360
18. Tang X, Pikal MJ (2004) Design of freeze-drying processes for pharmaceuticals: practical advice. Pharm Res 21:191–200
19. Velardi SA, Barresi AA (2008) Development of simplified models for the freeze-drying process and investigation of the optimal operating conditions. Chem Eng Res Des 86:9–22
20. Williams ML, Landel RF, Ferry JD (1955) The temperature dependence of relaxation mechanisms in amorphous polymers and other glass-forming liquids. J Am Chem Soc 77:3701–3707

Studies on Microwave-Vacuum Drying of Foods

Felipe Richter Reis

Abstract Microwave-vacuum drying is one of the most promising techniques for drying foods with proper quality at a relatively low cost. The short drying times make this process cost-effective thus allowing large scale dried food production. As shown in this chapter, microwave-vacuum dried foods include bananas, grapes, pomegranates, carrots, seeds, gels, honey and snacks. A couple of equations for calculating drying efficiency and modelling drying kinetics are presented. The values of effective moisture diffusivity for some other products are also presented. Most of the studies reported here investigate the effect of process conditions, especially vacuum pressure and microwave power level, on quality and engineering aspects of selected products. Some of them dealt with process optimization. The chapter presents studies from end the past century, when microwave-vacuum drying of foods started to be studied, until nowadays. The evolution of this technique can be confirmed by technical advances introduced over the years such as the use of product rotation, power intermittency, control of product temperature, osmotic pretreatment and improvements on the equipment design.

Keywords Microwave-vacuum drying · Microwave power level · Vacuum pressure · Process optimization

After a comprehensive survey in electronic databases, the first report found on microwave-vacuum drying of foods dates back to the end of the last century. Drouzas and Schubert [8] were among the first to study this issue. In their study, the influence of variable vacuum pressure and magnetron power level on drying rate and product quality aspects of bananas was investigated. The experimental apparatus was much like that described in Fig. 1.4 in chapter Introduction to Low Pressure Processes. The absolute pressure inside the vacuum vessel ranged from

F. Richter Reis (✉)
Food Technician Course, Instituto Federal do Paraná, Jacarezinho, Brazil
e-mail: felipe.reis@ifpr.edu.br

15 to 300 mbar. The power level varied between 150 and 850 W, being turned on for 10 s and then turned off for 20 s, repeatedly. Results showed that the changes in vacuum pressure did not affect the drying rate, while higher power level led to higher drying rate. Nevertheless, higher power level also caused product burning. In order to optimize the process toward product quality, a trial was conducted at 25 mbar and 150 W and the effects on product quality were monitored. As a result, a product presenting quality similar to a freeze-dried product was obtained, i.e., bright yellow color, excellent taste/smell and no shrinkage. In terms of stability, the final product obtained under optimal conditions presented water activity of about 0.35. Finally, the microwave-vacuum dried banana rehydration behavior was rather appropriate. In sum, this pioneer study showed that the microwave-vacuum drying technology was a promising way for obtaining high quality food products.

Drouzas et al. [9] studied the microwave-vacuum drying of a model gel imitating a concentrated orange juice. A microwave-atmospheric drying and a convective tunnel drying were carried out in parallel as reference. The effect of the sample position inside the microwave-oven, power level (640–710 W) and pressure (30–50 mbar) on the drying rate was elucidated. In addition, a sorption isotherm was generated and the color changes during the different drying processes were expressed in the L*a*b* color space. It was observed that a different position of the food material inside the microwave cavity leads to different drying rate. The power level positively affected the drying constant K (min^{-1}), which was used as another way to express the drying rate. On the other hand, an augment in pressure caused a decrease in K values and consequently, in the drying rates. The sorption isotherm generated presented a shape typical of high-sugar fruit materials. The changes in gel color were more pronounced for the microwave-atmospheric drying. The microwave-atmospheric dried product turned brown in the end of the process as expressed by decrease in lightness (L*), increase in redness (a*) and decrease in yellowness (b*).

Erle and Schubert [10] studied the effect of osmotic pretreatment as combined with microwave-vacuum drying on quality characteristics of apples and strawberries. The osmotic dehydration was carried out at the following conditions: temperature between 20 and 70 °C; osmotic solution sucrose concentration of 60 % (w/w) without calcium chloride, or 54 % of sucrose combined with 6 % calcium chloride (w/w). The microwave-vacuum drying apparatus consisted of a pilot-plant tunnel. The power level varied between 195 and 390 W and the time varied between 13 and 39 min. The absolute pressure inside the dryer was fixed at 5 kPa. The product quality was expressed as: sugar concentration, vitamin C content, calcium content, volume and visual appearance. Results showed that during the first 2 h of osmotic treatment the sucrose uptake by the strawberries reaches its maximum (25–30 % d.b.). The strawberries fructose and glucose content increased after 7 h of osmotic treatment, which was attributed to hydrolysis of sucrose. Both strawberries and apples retained almost 100 % of their original vitamin C content during the osmotic treatment, which decreased to about 60 % after the microwave-vacuum drying. Product volume was increased with the application of the microwave-vacuum drying, which was attributed to a pressure build-up inside the tissue as steam was created. In addition, the use of calcium chloride was believed to promote the formation of a gel with sucrose and pectin, increasing even more the product volume and generating a rigid and brittle texture that resembled a freeze-dried product. Finally, the use of electron microscopy showed that the structure of the osmotically pretreated/microwave-vacuum dried sample was more preserved at the cellular level than the sample that was only microwave-vacuum dried. Such results led to the conclusion that microwave-vacuum drying is a feasible way to obtain high quality dehydrated apples and strawberries. Furthermore, the final product quality can be enhanced by using an osmotic pretreatment.

Mousa and Farid [13] evaluated the changes in process efficiency, product temperature and product moisture content during the microwave-vacuum drying of banana slices. Process efficiency was expressed by two ways: drying efficiency and thermal efficiency, respectively shown below:

$$\eta_D = \frac{m_w \lambda_w}{W \Delta t} \times 100 \ \% \tag{1}$$

$$\eta_F = \frac{m_w \lambda_w + [M_w C p_w + M_s C p_s] \Delta T}{W \Delta t} \times 100 \ \% \tag{2}$$

where η_D is the drying efficiency (%); m_w is the mass of water evaporated during the time increment (kg), λ_w is the latent heat of vaporization of water (J kg^{-1}); W is the average microwave power (W); Δt is the time increment (300 s); η_F is the thermal efficiency (%); M_w and M_s are the masses of water and solids in the product (kg), respectively; Cp_w and Cp_s are the heat capacities of water and solids (J kg^{-1} K^{-1}), respectively; and ΔT is the product temperature variation during a specific heating period (°C). It was observed the drying efficiency was always smaller than the thermal efficiency, which was attributed to the significant contribution of the sensible heating during the drying process. The values of both efficiencies decreased during drying as a result of the water evaporation. In time: microwaves are well absorbed by water, whose content decreases during drying due to evaporation. The effect of vacuum on the process efficiencies was clearer at low moisture content of the product. In this case, the higher the vacuum used, the higher the process efficiency. Product temperature rose during drying in three different stages: the first rise took place from the initial product temperature to the boiling point of water at the respective pressure; the second stage consisted of a subtle rise corresponding to the majority of the free water evaporation; the third stage consisted of a sharp rise in temperature due to product heating in the end of the process. The rise in product temperature was more pronounced at higher pressures. The decrease in product moisture content was represented by a long constant drying rate period followed by a short falling drying rate period. In sum, as water is removed from the bananas, the drying efficiency decreases and the product temperature increases. Product temperature rise can be attenuated by increasing the level of vacuum.

Clary et al. [1] performed the vacuum-drying of grapes at fixed and decremental magnetron power levels, ultimately finding optimum process conditions that led to high quality dried grapes. The power levels tested ranged from 0.5 to 1.5 kW for the fixed microwave tests and from 3.0 to 0.5 kW for the decremental power level tests. A pressure of 2.7 kPa inside the drying vessel was used in all cases. A parameter called specific energy, which expresses the microwave power applied in each trial (W h/g of fresh grapes), was measured and calculated. Results showed that the fixed microwave tests yielded products either burned—at high power levels—or too moist—at low power level. On the other hand, the use of decremental power levels during drying led to grapes with better quality. This happened because the power reduction during drying avoided product overheating. A multiple regression analysis showed that the specific energy was responsible for most of the variation in the final moisture content of the grapes. The changes in product temperature during drying were also monitored. As observed in previous studies, the product temperature variation consisted of: (1) an initial increase from the ambient temperature to the water boiling temperature at the respective pressure; (2) a temperature plateau which represented a balance between heating and drying; and (3) a product heating phase. The optimum specific energy was 0.1 W h/g in stage 1; 0.575 W h/g in stage 2; and 0.245 W h/g in stage 3, totalizing 0.92 W h/g in the whole drying process. When the optimum drying conditions were applied, grapes with low moisture content and puffed/crispy character were obtained.

Cui et al. [5] studied the changes in product temperature during the microwave-vacuum drying of carrot slices. In addition, the drying kinetics under different process conditions was elucidated and modeled. The magnetron power level varied roughly between 290 and 359 W. The slices thickness was either of 4, 8 or 10 mm. The vacuum pressure was fixed at 30, 51 or 71 mbar. Results showed that the temperature is the same in the whole carrot slice when its thickness is lower than 8 mm, due to negligible internal mass transfer resistance. In addition, the temperature profile graphs showed that the temperature changes in the product during drying can be divided in three stages: warming-up, water vaporization and heating-up, as observed in previous studies. The drying curves could be divided in an initial constant rate period followed by a falling rate period. A theoretical model based on initial moisture content, energy absorbed by sample, mass of sample and latent heat of evaporation of water was used to fit the experimental data (see [4] for more details). The model was suitable for the constant rate period. For modeling the falling rate period, a correction factor had to be introduced. The correction factor value was obtained by non-linear regression. In conclusion, temperature in carrot slices is uniform since the slice is not too thick. Temperature changes during drying followed the pattern usually observed for microwave-vacuum drying. Product moisture content and temperature during drying were well predicted by the developed empirical models.

Clary et al. [3] performed an interesting study where the microwave-vacuum drying of grapes was carried out under controlled microwave power. The power control was based on the measurement of product temperature. In addition, the effect of several process and product parameters on the grapes quality was evaluated. Furthermore, nutritional characteristics of the microwave-vacuum dried grapes were compared to fresh grapes and sun-dried raisins. A sensor was used to acquire the grapes temperature, which was sent to a controller that immediately acted on the magnetron, controlling the microwave power generated. When the temperature in the product surface approached the predefined maximum temperature, the power generation was automatically reduced. Power level ranged from 0 to 3 kW and pressure was fixed at 2.7 kPa. Those authors observed that product temperature is a proper parameter to be used for controlling the microwave power level during the process. In time: product burning, which is a common drawback during the microwave drying of foods, was avoided. This behavior could be confirmed visually and by plots of product surface temperature versus drying time. With regard to the effect of product temperature, initial sugar/moisture content, drying time and specific energy on the moisture content and puffed character of the dried grapes, it was shown that the influence of product temperature was the most significant. Surface plots allowed the optimization of the microwave-drying process toward low final moisture content and high incidence of puffed grapes. In this sense, a temperature of about 70 °C was considered the most appropriate. With respect to the nutritional profile of the microwave-vacuum dried grapes, two aspects should be highlighted: the product presented around 4.5 times the content of non-heat sensitive elements of the fresh fruit; and heat sensitive nutrients (vitamins) were better preserved in the product when compared to sun-dried raisins. Summarizing, the control of power level on the basis of product temperature is valid for preventing grapes burning. Product temperature seems to control the quality of the final product. Finally, the microwave-vacuum dried grapes were distinctively nutritious.

Jaya and Durance [11] built the microwave-vacuum drying curves of soft solid alginate-starch gels for different microwave power levels. The kinetics were tentatively fitted by two kinds of exponential models: one of them, more typical, related moisture content to drying time and the other one, more innovative, related moisture content to microwave power level. The dryer comprised a rotating transparent polyethylene drum located inside a vacuum chamber operating at 25 mmHg and supplied with microwave energy (300–1100 W). The parameters of the drying models were obtained on the basis of the plots of moisture content as a function of time or as a function of microwave energy. In addition, linear relationships between the parameters of the models and time or power level were successfully developed. Regarding the effect of power level on drying rates, it was shown that the use of higher power yielded higher drying rates. Consequently, at a given time of drying, lower moisture contents were detected in the samples subjected to higher power levels. Finally, the proposed models were found to be suitable to fit the drying data, yielding values of predicted moisture content that were very close to the experimental data. Summarizing, the drying curves of microwave-vacuum dried hydrogels were successfully obtained and modeled by exponential models and the drying rates showed to be power level-dependent.

Cui et al. [6] optimized the microwave-vacuum drying of honey and measured the color and the concentration of sugars and flavors in the final product. Variable levels of three parameters, viz. microwave power, vacuum pressure and sample thickness, were used. The first parameter varied between 290 and 330 W; the second parameter ranged between 30 and 50 mbar; and the third parameter varied from 8 to 16 mm. Their effect on the product core and surface temperature and drying rate was evaluated. Results showed that the product core and surface temperature were similar for low sample thickness (8 mm or less). As thickness increased, a temperature gradient developed along the axial direction of the cylindrical sample. The effect of the power level on the temperature of 8 mm thick samples was more significant than the effect of vacuum pressure. With regard to drying rate, such parameter decreased with sample thickness and increased with power level. Product color, as expressed in the CIE L*a*b* scale, changed negligibly during drying. Fructose and glucose content barely increased, while sucrose and maltose barely decreased during drying. Such behavior was attributed to the activation of the enzyme invertase in the honey on sucrose and maltose. Even though some of the acids and esters were probably decomposed into alcohols, aldehydes and ketones, the original honey flavor was preserved during drying. Therefore, microwave-vacuum drying showed suitable for obtaining dry honey of proper quality with low drying time.

Therdthai et al. [15] combined osmotic dehydration and microwave-vacuum drying for extending the shelf-life of mandarins. The osmotic solution was composed by water and glycerol in the proportions of 9:1, 7:3 and 5:5 (w/w). The effect of the osmotic process on the dielectric properties of mandarins was evaluated. A dielectric measurement kit was used for this purpose. The drying was carried out at two different microwave power levels: 960 and 1,280 W. The frequency of operation was 2.45 GHz and the vacuum pressure was 13.33 kPa. The drying kinetics was generated and several thin-layer drying models were used to fit the data. In addition, quality parameters of the final product were measured. Those authors observed that the osmotic dehydration promoted changes in the mandarins' dielectric properties. The dielectric constant, which is the ability of a material for storing electrical energy from an external electric field, suffered a decrease. The loss factor, which is the ability of one material to convert electrical energy into thermal energy, increased. Consequently, the loss tangent, which is the ratio between the loss factor and the dielectric constant, also increased. Finally, the penetration depth, which is the distance inside the product where the microwave power drops to 36.8 % of the transmitted value on the surface, suffered a decrease. All these effects were more pronounced as the glycerol concentration in the osmotic solution increased. Increase in the microwave power level promoted increase in the drying rates. In addition, increase in glycerol level also promoted increase in drying rates. Such behavior was attributed to the increase in loss factor after pretreating the mandarins at high glycerol concentrations. The drying kinetics was best fitted by the Page model. Such model provided a fit with the lowest error and the highest correlation coefficient. Changing the sucrose to glycerol ratio affected color, β-carotene content and texture of the final product. More specifically, increase in the glycerol concentration promoted decrease in lightness, colorfulness and hue angle. In this sense, the authors concluded that the color of the mandarins was more intense when higher levels of sucrose were used in the osmotic pretreatment. The same behavior was observed for β-carotene content, i.e., high levels of sucrose preserved this pigment in a better way. The authors suggested that the results for color and β-carotene content are correlated. The β-carotene was also affected by different power levels. The use of a higher power level yielded mandarins with lower β-carotene content. Finally, the hardness of the mandarins was affected by different sucrose to glycerol ratios, but not by different power levels. In this regard, the greater the proportion of glycerol in the osmotic solution, the softer the final product obtained. Concluding, the use of an osmotic pretreatment prior to microwave-vacuum drying of mandarins improved the drying rate and, under specific conditions of pretreatment, color, β-carotene content and texture observed in the dried product were rather appropriate.

Tian et al. [16] investigated the effect of variable microwave-vacuum drying conditions on quality features of lotus seeds (*Nelumbo nucifera*). The process was optimized by using the response surface methodology and the desirability function. In addition, selected quality aspects of the microwave-vacuum dried lotus seeds were compared to those of hot air dried ones. A central composite design was used to evaluate the effect of microwave power (2.0–4.0 kW), vacuum degree (−70.0 to −90.0 kPa) and power on/off ratio (68/52–99/21 s) on drying time, shrinkage ratio, rehydration ratio and whiteness index of the lotus seeds. Second order polynomial models successfully explained the effect of the independent variables on the responses. Such conclusion was made on the basis of the significance of the models, insignificance of the "lack of fit" term and high values of determination coefficient ($R^2 \geq 0.91$). Response surface and contour plots generated on the basis of the empirical models showed that: increase in power level, vacuum degree and on/off ratio led to lower drying time and lower shrinkage; intermediate levels of vacuum combined with high power level and high on/off ratio yielded seeds with high rehydration ratio; roughly speaking, intermediate levels of power, vacuum and on/off ratio, when combined, led to high whiteness on the final product. The optimum conditions of drying as obtained by the desirability function were: 3.2 kW of power, −83.0 kPa of vacuum and an on/off period of 96/24 s. According to the model, when such conditions are used, a short drying time of 9.5 min and an ideal quality, viz. shrinkage ratio of 38.52 %, rehydration ratio of 155.0 % and whiteness index of 67.67, were obtained. The optimum process conditions were tested in three further trials and the generated results showed that the predicted and the experimental values were very close. Finally, when comparing the microstructure of the microwave-vacuum dried lotus seeds to that of the hot air dried ones, the former presented more pores than the latter, which ultimately led to better rehydration. Furthermore, the appearance of the microwave-vacuum dried lotus seeds was more attractive than that obtained by hot air drying. In conclusion, when applied under optimized conditions, microwave-vacuum drying showed suitable for obtaining high quality dried lotus seeds.

Kraus et al. [12] studied the effect of variable microwave-vacuum drying conditions on the drying kinetics, effective diffusivity and volume of starch-based snack pellets. In addition, the changes in product dielectric properties during drying were monitored. The drying kinetics was tentatively fitted with thirteen thin-layer drying models. The most suitable model was found to be the Balbay and Sahin (Balbay and Sahin 1) model, shown below:

$$M_R = (1 - a) \exp\left(-kt^n\right) + b \tag{3}$$

where M_R is the dimensionless moisture ratio presented in Eq. 2 in chapter Studies on Conventional Vacuum Drying of Foods; a, b, k and n are model constants and t is time (min). This model provided a fit with the highest values of coefficient of determination (R^2) and the lowest values of reduced chi-square. Among the independent variables chosen, it was observed that: increase in microwave power level (400–800 W) led to increase in drying rates; increase in sample amount (100–300 g) led to decrease in drying rates; changes in vacuum pressure (20–100 mbar) barely affected the drying rates. Since the product form changes from a cylinder to a sphere during drying, effective diffusivities were calculated by using Fick's second law for both cylindrical and spherical pellets. The difference in calculated effective diffusivities between these two forms was insignificant, though. On the other hand, effective diffusivity values were affected by different process conditions. Effective diffusivity increased with increasing microwave power level and decreasing sample amount, while vacuum pressure did not affect this property. Expanded volume, a desired feature in extruded snacks, was obtained when high microwave power levels were used. Finally, the product dielectric properties decreased during drying. This behavior was attributed to a reduction in the capacity of a material to absorb microwave energy and convert it into heat while water is being removed from it. Summarizing, the Balba and Sahin model provided the best fit to the kinetic data during drying of starch-based pellets. Changes in drying conditions affected the drying rates and the effective diffusivities, except from vacuum pressure. The best quality was observed in the pellets dried under high power levels.

Patel et al. [14] optimized the fermentation and the microwave-vacuum drying of an Indian vegetable-based food named Dhokla. The response surface methodology and the desirability function were employed as optimization techniques for both processes. While the fermented Dhokla contains 65–70 % moisture, the dried product (instant Dhokla mix) contains 5–6 % moisture. The product was formulated with rice, Bengal gram, black gram, water and salt. Before fermentation, the ingredients were hydrated and ground. After fermentation, the product was separated in two lots. One of them was cooked and the other was dried. For optimizing the fermentation, variable levels of time, temperature, moisture content and rice to Bengal gram ratio were used. Total acidity, lactic bacteria count, firmness and sensory acceptability were used as responses. The objective was to obtain maximum values for all responses simultaneously. Then, part of the fermented batter was microwave-vacuum dried. The fermentation conditions used were the optimum, i.e. time of 12.5 h, temperature of 26.6 °C, moisture content of batter of 65 % w.b. and rice to Bengal gram ratio of 1.2. The effect of variable drying conditions on quality parameters of the final product was evaluated. The independent variables chosen were: thickness of batter (10–17 mm), microwave power density (3.5–10.0 W g^{-1}) and pulsating ratio (1.3–2.0), that was the total drying time divided by power-on time. The responses studied were: bulk density, rehydration ratio, color and sensory acceptability. The vacuum in the drying cavity was set to −80 kPa. A plate rotating at 17 rpm was used to support a glass bowl containing the sample. Weight measurements were performed throughout the drying process until the product achieved a moisture content of about 4–6 % d.b. Results expressed by perturbation graphs show that, among the independent variables tested: product bulk density was specially affected by thickness of batter; rehydration ratio was mostly affected by pulsating ratio; total color change and sensory acceptability were sensitively affected by thickness of batter and power density. The optimized drying conditions obtained were: thickness of batter of 17 mm, microwave power density of 10 W g^{-1} and pulsating ratio of 1.3. When using such conditions, an optimum quality was observed in the final product, viz. bulk density of 1,014.22 kg m^{-3}, rehydration ratio of 4.55, total color change of 9.57 and acceptability score of 6.88. Additional results showed that: a low value of calculated specific energy consumption was associated with the microwave-vacuum drying process; increase in thickness and pulsating ratio and decrease in power density led to low drying rates; the Page model provided a proper fit for the experimental drying data; effective diffusivity values as calculated by the Fick's second law of diffusion for an infinite slab ranged between 6.89×10^{-8} and 1.10×10^{-7} m^2 s^{-1}.

The changes in effective moisture diffusivity during microwave-vacuum drying of pomegranate arils under different conditions were studied by Dak and Pareek (7). The drying process was conducted at a power level ranging from 25 to 95 W, an absolute pressure ranging from 25 to 195 mmHg and a sample mass varying between 65 and 235 g. Effective moisture diffusivity was calculated by the Fick's second law of diffusion for a cylinder. Results showed that the higher the power level, the higher the effective moisture diffusivity. In addition, the lower the sample mass, the higher the effective moisture diffusivity. On the other hand, vacuum pressure did not exert significant effect on effective moisture diffusivity. The values of effective moisture diffusivity varied between 5.18×10^{-11} and 6.58×10^{-10} m^2 s^{-1}. The relationship between effective diffusivity and moisture content was well described by a third order polynomial model. The effect of power level and sample mass on the values of effective diffusivity was well modelled ($R^2 = 0.99$) by a second order polynomial equation. In this sense, it was concluded that the effective moisture diffusivity of pomegranate arils is significantly affected by different microwave-vacuum drying conditions.

References

1. Balbay A, Sahin O (2012) Microwave drying kinetics of a thin-layer liquorice root. Dry Technol 30:859–864
2. Clary CD, Wang S, Petrucci VE (2005) Fixed and incremental levels of microwave power application on drying grapes under vacuum. J Food Sci 70:E344–E349
3. Clary CD, Mejia-Meza E, Wang S et al (2007) Improving grape quality using microwave vacuum drying associated with temperature control. J Food Sci 72:E23–E28
4. Cui Z-W, Xu S-Y, Sun D-W (2004) Microwave-vacuum drying kinetics of carrot slices. J Food Eng 65:157–164
5. Cui Z-W, Xu S-Y, Sun D-W et al (2005) Temperature changes during microwave-vacuum drying of sliced carrots. Dry Technol 23:1057–1074

6. Cui Z-W, Sun L-J, Chen W et al (2008) Preparation of dry honey by microwave-vacuum drying. J Food Eng 84:582–590
7. Dak M, Pareek NK (2014) Effective moisture diffusivity of pomegranate arils under going microwave-vacuum drying. J Food Eng 122:117–121
8. Drouzas AE, Schubert H (1996) Microwave application in vacuum drying of fruits. J Food Eng 28:203–209
9. Drouzas AE, Tsami E, Saravacos GD (1999) Microwave/vacuum drying of model fruit gels. J Food Eng 39:117–122
10. Erle U, Schubert H (2001) Combined osmotic and microwave-vacuum dehydration of apples and strawberries. J Food Eng 49:193–199
11. Jaya S, Durance TD (2007) Effect of microwave energy on vacuum drying kinetics of alginate-starch gel. Dry Technol 25:2005–2009
12. Kraus S, Sólyom K, Schuchmann HP et al (2013) Drying kinetics and expansion of non-predried extruded starch-based pellets during microwave vacuum processing. J Food Process Eng 36:763–773
13. Mousa N, Farid M (2002) Microwave vacuum drying of banana slices. Dry Technol 20:2055–2066
14. Patel DN, Sutar PP, Sutar N (2013) Development of instant fermented cereal-legume mix using pulsed microwave vacuum drying. Dry Technol 31:314–328
15. Therdthai N, Zhou W, Pattanapa K (2011) Microwave vacuum drying of osmotically dehydrated mandarin cv. (*Sai-Namphaung*). Int J Food Sci Tech 46:2401–2407
16. Tian Y, Zhang Y, Zeng S et al (2012) Optimization of microwave vacuum drying of lotus (*Nelumbo nucifera* Gaertn.) seeds by response surface methodology. Food Sci Technol Int 18:477–488

Comparative Studies

Felipe Richter Reis

Abstract This chapter is the longest and probably the most important for the decision-making between the available drying techniques. Several vacuum drying techniques are compared to each other and to conventional drying techniques in terms of product quality and engineering aspects. Vacuum drying usually provided better results than conventional processes, such as hot air drying, in terms of product quality. Nevertheless, hot air drying is still cheaper than vacuum drying and depending on the product value it could be more interesting to use a cheaper drying technique. The differences between conventional vacuum drying, freeze-drying and microwave-vacuum drying are elucidated and quantified. New processes of subatmospheric drying that have been developed during the last years are also presented. A search in electronic databases shows that vacuum and conventional drying processes for foods have been compared since the 1940s and this fashion of research remains popular until nowadays. Definitely, this chapter is very useful for entrepreneurs and researchers interested in choosing the most suitable drying technique for a selected food product.

Keywords Vacuum drying · Freeze-drying · Microwave-vacuum drying · Hot air drying

Dunlap Jr. [10] built a vacuum drying apparatus for drying blocks of carrot and potato pieces. Blocks of pieces were a common way of disposing vegetables that were sent to the battle field during wars. The vegetables were cut either in the form of strips or cubes and compressed in blocks measuring approximately $3 \times 2 \times 1$ in. The typical process for preparing potato blocks consisted in drying the potato pieces to 7 % moisture and compressing them. Nevertheless, this process caused the crushing of potatoes. In order to avoid crushing, the author

F. Richter Reis (✉)
Food Technician Course, Instituto Federal do Paraná, Jacarezinho, Brazil
e-mail: felipe.reis@ifpr.edu.br

© The Author(s) 2014
F. Richter Reis (ed.), *Vacuum Drying for Extending Food Shelf-Life*, SpringerBriefs in Applied Sciences and Technology, DOI 10.1007/978-3-319-08207-3_5

performed the drying of potatoes to 12–15 % moisture followed by compression and rapid drying to 7 % moisture. The new method was tried under two circumstances: vacuum-oven or radio-frequency heating in vacuum. The effects of these two procedures on drying rates and product quality were compared. During the drying process, pressure varied between 10 and 50 mmHg and temperature was either 70 ± 1 °C for the vacuum-oven drying or 60 ± 1 °C for the radio-frequency heating in vacuum. It was observed that the use of radio-frequency heating led to higher drying rates. The final product quality, expressed as perceived color, taste, texture, flavor and measured rehydration capacity, was better when conventional vacuum drying was used, though. Discoloration by browning was shown to be higher when radio-frequency heating was used. This behavior was attributed to the combination of high temperatures and high moisture content in the product during radio-frequency heating. In conclusion, the conventional vacuum drying was considered by the author as the best one among the processes studied.

Lin et al. [21] performed parallel trials of hot air drying (AD), freeze-drying (FD) and microwave-vacuum drying (MVD) of carrot slices. The first was performed at 70 °C. The second was carried out at 1.6 mmHg of pressure and temperatures of 20 °C at the heating plate and −55 °C at the condenser. The third was performed at 100 mmHg and microwave power of 3 kW (first 19 min), 1 kW (subsequent 4 min) and finally 0.5 kW (last 10 min). The effect of the three processes on the quality of the final product was studied. With regard to drying rates, MVD was more efficient than AD, which was more efficient than FD. It took 33 min to dry 1 kg of carrot to a final moisture content of 10 % (w.b.) by MVD, while 8 h were necessary for the AD and 72 h for the FD process to promote the same effect. The density of the final product was higher for the product dried by AD, followed by MVD and FD. Such results correlated inversely to the rehydration capacity, which was confirmed to be higher for less dense products. In addition, rehydration rates were found to be higher at the beginning of rehydration and at higher temperatures. The dried and the rehydrated carrot slices texture as measured instrumentally was different for the different processes tested. The product dried by AD was harder, probably due to a case hardening effect, which was not observed in the carrots dried by MVD and FD. With respect to the rehydrated product, carrots dried by FD were the softest ones, while those dried by AD were the hardest. With regard to product color, carrots dried by FD were the lightest. On the other hand, the rehydrated product color was more similar for the three processes studied. Roughly speaking of carotenoid and vitamin retention, the FD process was the best and the AD process was the worst. The dried and the rehydrated carrots were also evaluated by a sensory panel. In this regard, the product dried by MVD received significantly higher ratings for texture and overall acceptability, and similar ratings for color and aroma/flavor when compared to that dried by FD. On the other hand, carrots dried by AD presented a significantly lower sensory quality. Nevertheless, the rehydrated product obtained by AD presented an improved quality. The higher rating of aroma/flavor of the rehydrated carrots obtained by MVD when compared to those obtained by FD reinforces the hypothesis that FD promotes significant loss of volatile compounds. In sum, the new process being tested in that study, namely MVD, was considered suitable for obtaining dried carrot slices with proper quality at a relatively low cost. In addition, the unique puffed structure obtained by MVD is considered highly desirable in snack foods.

Krokida et al. [18] investigated the rheological phenomena behind the rehydration of dehydrated vegetable foods. Apple, banana, carrot and potato cylinders were dehydrated and then rehydrated. The drying methods used were: air drying, vacuum drying, freeze drying and osmotic dehydration/freeze drying. The drying conditions were, respectively: 70 °C and 7 % air relative humidity; 70 °C and 33 mbar; initial sample temperature of about −45 °C and 0.04 mbar of pressure; and immersion in a 50 % sucrose solution at 40 °C for 10 h/same freeze drying conditions. Rehydration was performed by using air at 50 °C and 80 % of humidity. The rheological properties measured were: maximum stress, maximum strain, elastic parameter and viscoelastic exponent. The changes in the values of the rheological parameters during dehydration and rehydration were monitored by recording the results of compression tests. By plotting stress versus strain for freeze-dried apple at constant moisture content, those authors observed that the curves for dehydration and rehydration do not coincide. This finding led to the conclusion that the dehydration procedure is irreversible. Maximum stress (kPa) increased with decreasing moisture content during both dehydration and rehydration. The phenomenon of hysteresis was observed for all the drying processes, i.e., the changes in maximum stress during drying did not coincide with those observed during rehydration. In fact, the maximum stress for rehydration dropped in comparison to that for dehydration. The hysteresis was pronounced for the freeze-dried product and subtle for air and vacuum dried products. Nevertheless, osmotic dehydration seems to have reduced the hysteresis observed for the freeze-dried product. Similar changes were observed for maximum strain with varying moisture content. As a result, freeze-drying appears to generate more fragile final products which might become stronger, though, by the incorporation of solids during the osmotic pretreatment. On the other hand, air and vacuum-dried products present a tougher structure resultant from the shrinkage effect. It was observed a decrease in the products' elastic parameter with decrease in moisture content, except from the freeze-dried product. At the same time, increase in viscoelastic exponent with decrease in moisture content was observed for almost all products. Once again the freeze-dried product behaved inversely. Such results led to the conclusion that most of the drying techniques studied yield fruits and vegetables with high elasticity and low viscosity after rehydration. On the other hand, the freeze-dried fruits and vegetables become less elastic and more viscous when rehydrated. Consequently, rehydrated air and vacuum dried products present a viscoelastic behavior similar to that of fresh products. Despite the superior quality of freeze-dried products, when rehydrated, they do not recover the viscoelastic features of fresh products.

Tsami et al. [38] evaluated the influence of different drying techniques on the sorption properties of food powders. The food sample comprised a gel composed by water, sugars, pectin and citric acid. The gel presented 60.8 % d.b. of initial moisture content. Once solidified, the gel was submitted to hot air, vacuum, freeze or microwave drying. Hot air drying was carried out at 70 °C for 36 h after a pretreatment with microwave at 595 W for 1.5 min. Vacuum drying was performed at 70 °C and pressure of 25 mbar for 16 h. Freeze-drying was performed at a condenser temperature of 30 °C and pressure of 0.04 mbar for 72 h. Microwave drying was carried out intermittently (on/off period of 10–15 s/15 s) at atmospheric pressure and 595 W of power for 30 min. The final moisture content of the gel ranged between 3–6 % d.b., depending on the drying method. The dried gel was ground to about 1 mm. The sorption behavior of the powders was monitored hygroscopically by using specific equipment operating at 25 °C. Besides sorption characteristics, the powder was assessed for densities, color and porosity. The shape of the isotherms was observed to be sigmoid. As expected for this type of isotherm, small amounts of water were sorbed at low water activity and large amounts of water were sorbed at high water activity. This behavior is typical of high sugar foods. Among the equations used for fitting the sorption data, viz. GAB, Hasley and Oswin models, the former was found to be the best. The GAB model [24] is shown below:

$$X = X_m C k a_w / (1 - k a_w)(1 - k a_w + C k a_w) \tag{1}$$

where X is the moisture content of the material (kg water kg dry solid^{-1}); a_w is the water activity; X_m is the monolayer moisture content (kg water kg dry solid^{-1}); and C and k are model constants. Comparing the various drying techniques, the highest adsorptive capacity was observed in the freeze-dried product, followed by the microwave-dried, the vacuum-dried and the air-dried product. This result was attributed to the high amount of small sized pores in the freeze-dried product. This microporous structure is generated by rapid freezing. The microwave-dried product possesses a porous structure due to rapid evaporation of water within the food. The vacuum-dried product also presents a porous structure, but the presence of fewer and larger pores leads to lower sorption capacity. Finally, the air-dried product presents comparatively the lowest number of small pores. This ultimately led to the worse sorption capacity. Results obtained on the basis of the isotherms could be confirmed by the bulk density and the bulk porosity measurements. In general terms, product color was better preserved by the methods using vacuum. In conclusion, results suggested that the freeze-dried product presented the best quality, but the microwave-dried product presented a porous structure and the vacuum-dried product presented a good color. These findings were an indicium that the combination of microwave and vacuum could be an interesting option. Studies on microwave-vacuum drying were starting at that time, coincidently.

Krokida et al. [19] investigated the effect of the method of drying on color of selected fruits and vegetables. Apples, bananas, carrots and potatoes were cut into cylinders. The pieces were subsequently treated by: hot air drying (alone and combined with microwave or osmotic pretreatment), vacuum drying and freeze-drying. Hot air drying was carried out at 70 °C with 7 % air-relative humidity. As mentioned before, some trials comprised a pretreatment before hot air drying: (1) microwave operating at 810 W for 1 min; or (2) osmotic dehydration at 40 °C and 40 % of sucrose in the solution for 10 h. Vacuum drying was performed 70 °C and 33 mbar of pressure. Freeze-drying was performed under a pressure of 0.04 mbar and at a temperature not informed. Color of the samples was monitored throughout the drying processes. The CIE L*a*b* scale was used to express the results. The color kinetics was tentatively fitted by a first order kinetic model. The goodness of fit was evaluated by computing the values of the residual sum of squares. Those authors found out that the mathematical model used was proper for fitting the experimental data, as denoted by low values of residual sum of squares. The changes in lightness (L*) followed an unclear pattern for some methods and foods, which was attributed to significant experimental error. For others, L* clearly changed toward the black color. In addition, negligible changes in L* of osmotically pretreated/convectively dried and freeze-dried foods were detected. With regard to samples' redness (a*) and yellowness (b*), increase was observed throughout the drying processes. Although, such augment was barely detectable in osmotically pretreated/convectively dried and freeze-dried materials. According to those authors, the increase in a* and b* values is consequence of browning reactions. This finding led to the conclusion that osmotically pretreated/convectively dried and freeze-dried materials suffered less browning than those dried by other methods.

Maache-Rezzoug et al. [23] developed a new drying process named "Déshydratation par Détentes Successives"—DDS, i.e., dehydration by successive pressure drops. The DDS method was applied to the polysaccharide scleroglucan, a glucose polymer, in the form of a coagulum. Two conventional drying methods viz. air drying and vacuum drying were conducted in parallel. The effect of various drying conditions on drying kinetics and final product hydration capacity was evaluated. The DDS method consisted in placing the sample in an autoclave connected to a vacuum chamber. The desired vacuum was attained and the valve placed between the autoclave and the chamber was closed. Then, air was introduced until high pressure was attained in the autoclave. After a few seconds, the valve was opened, generating a pressure drop. This procedure was repeated until the desired moisture content was reached. The variable drying conditions during the DDS were: processing time (5, 8 or 10 s), processing pressure (4.5, 6.0 or 7.5 bar) duration of one cycle (28, 31 or 33 s) and number of pressure drops per hour (109, 116 or 128). The pressure during the vacuum phase of the DDS was fixed at 0.05 bar and the duration of this phase was fixed at 20 s. The air drying was conducted at different air flow rates (300, 500 or 700 l/h) and oven temperatures (30, 70 or 80 °C). The vacuum drying was carried out at various processing pressures (0.05, 0.10 or 0.20 bar) and heating plate temperatures (20 or 50 °C). After drying, the sample was milled and solubilized in water with the blades of a viscometer. The generated maximum torque was used to express the degree of hydration of the scleroglucan powder. The final product microstructure was evaluated by generating micrographs with an electronic microscope. The drying kinetics of the DDS was affected by processing time, but not by processing pressure. The DDS-dried product hydration capacity increased with decrease in drying time, decrease in autoclave pressure/temperature and increase in the number of pressure drops per hour. The drying kinetics of the air drying was significantly influenced by air temperature, while air flow rate presented little effect. The air dried product hydration capacity was affected by air temperature, where the higher the temperature, the better the hydration. With regard to vacuum drying, it was observed that when the pressure and temperature were combined as to allow the boiling of water inside the product (0.1 bar and 50 °C), the drying was faster and consequently the product hydration capacity was higher. The generated micrographs showed that the DDS-dried scleroglucan presents the most porous structure, followed by the vacuum-dried scleroglucan and the air-dried one. Such highly porous structure ultimately leads to better hydration, confirming the results of the rheological analyses. To sum up, the DDS method showed valid for producing dried scleroglucan with proper hydration capacity, while vacuum drying once again showed superior to air drying in terms of product quality.

Martínez-Soto et al. [25] investigated how different pretreatments and drying methods affected the drying kinetics and the quality of oyster mushrooms (*Pleutorus ostreatus*). Before drying, the mushrooms were treated by blanching (80 °C/3 min), immersion in sodium metabisulphite (1 or 5 g/L/10 min) or in citric acid (1 or 5 g/L/10 min). Then, the mushrooms were dried by three methods: hot air-drying (65 °C/1.5 m/s), vacuum-drying (55 °C/1,334 Pa) or freeze-drying (0 °C at heating plate/−55 °C at condenser/7 Pa). The quality of the final product was measured as per rehydration capacity, color and sensory evaluations. Fresh mushrooms were used as control. The drying kinetics showed that the constant rate period was short. Consequently, the drying of mushrooms is controlled by the diffusion during the falling rate period. Drying rates were higher for hot air-drying, followed by vacuum-drying and freeze-drying. As expected, drying rates decreased with the advance of drying. The use of blanching pretreatment increased the drying rates. Freeze-dried samples presented the highest rehydration capacity, while vacuum-dried ones presented the lowest rehydration capacity. These results were attributed to low and high shrinkage during drying, respectively. Blanching negatively affected the rehydration process. Regarding color, freeze-dried mushrooms were clearly lighter and more attractive than those obtained by air- and vacuum-drying. Here, once again blanching led to poor quality, i.e., a dark final product. The sensory evaluation showed that the products with better appearance were: mushrooms treated with 1 g/l sodium metabisulphite and air-dried; mushrooms treated with 1 g/l sodium metabisulphite and vacuum-dried; and mushrooms treated with 5 g/l sodium metabisulphite and freeze-dried. These products were submitted to another sensory test, where flavor, texture and overall acceptability were measured. Results showed that these sensory attributes are similar for air-dried, freeze-dried and fresh mushrooms, while the vacuum-dried product presented poor flavor. The authors concluded that freeze-drying yields mushrooms of better quality in some regards, but its high cost must be taken into account.

Grabowski et al. [12] tested various drying treatments subsequent to an osmotic pretreatment for improving the shelf life of cranberries (*Vaccinium macrocarpon*). The quality of the final products and the energy requirements of each process were used to choose the best drying technique. The osmotic pretreatment consisted of immersion of cranberry halves in a 67.5 °Brix sucrose solution at 50 °C for 5 h. This treatment reduced the moisture content of fresh cranberries from 80 to

50 % w.b. Subsequently, hot air drying, vacuum drying, freeze-drying, conventional fluid bed drying, vibrated fluid bed drying and pulsed fluid bed drying were used to dry the cranberries. The temperature was set at 90 °C for all the drying processes, except from freeze-drying, conducted at 30 °C and 270 Pa. The vacuum-drying was carried out at 20 kPa. The convective processes used air velocities of 0.9 m/s (hot air), 1.4 m/s (pulsed fluid bed and vibrated fluid bed) and 1.8 m/s (conventional fluid bed). The cranberries' layer varied from 40 mm (vacuum dryer) to 200 mm (vibrated fluid bed dryer). The quality indices evaluated in the dried cranberries were: water activity, anthocyanins content, rehydration capacity, color and taste. The energy efficiency of the selected drying processes was computed on the basis of the instantaneous and the cumulative energy efficiencies. While the former is the ratio between the energy used for evaporation and the input energy at time t, the latter is the integration of the former over a given time interval. The drying kinetics pattern observed for all convective drying equipment was similar, i.e., an initial drying period, followed by a constant rate period and the falling rate period. Reducing the height of the cranberry bed from 20 to 10 cm or increasing the temperature from 90 to 100 °C promoted a remarkable increase in drying rate during the convective drying processes. One of the objectives of that research was to find out the effect of osmotic dehydration on subsequent drying kinetics. In this regard, it was observed that the osmotic treatment reduces the drying rate. Nevertheless, the non-thermal water removal promoted by osmotic dehydration offsets the slow-down effect of the infused sugar on the drying rate. Therefore, from the economical viewpoint, the use of osmotic pretreatment is advantageous. With respect to product quality, the freeze-dried product presented the best taste, color and rehydration capacity, while anthocyanins' content was similar to that of the products dried by other methods. Apart from freeze-drying, the quality indices of the products processed in all other dryers were similar to each other and acceptable by consumers. Regarding energy performance, the instantaneous energy efficiency decreased during drying due to the progressive decrease in heat and mass transfer. The cumulative energy efficiency varied for the different dryers. In this sense, the fluid bed-type dryers showed more energy efficient than the others, especially the vibrated fluid bed dryer and the pulsed fluid bed dryer. In sum, even though the freeze-dried product presents superior quality, the products dried in fluid bed type-dryers also present acceptable quality with the advantage of being processed with superior energy efficiency.

Rahman et al. [30] studied the characteristics of the pores of dried tuna fish as obtained by different drying processes. Pores affect the physical properties of dehydrated foods. Therefore, knowing properties such as porosity, density and pore size is important for designing food drying processes. The drying techniques studied were: hot air-drying (70 °C, 3.4 % air relative humidity, 1 m/s air velocity); vacuum drying (70 °C, 2 kPa) and freeze-drying (−20 °C heating plate temperature, −65 °C condenser temperature, 108 Pa). Fresh tuna fish was evaluated for proximate composition, total microbial count, freezing point, several types of density and porosity. Besides these analyses, the dried products were also evaluated for wetting angle, pore size distribution at low and high pressure, color and peroxide value. The apparent density was greater for the fresh tuna, followed by the air-dried product, the vacuum-dried product and the freeze-dried product, suggesting that the latter presents larger pores. With concern to porosity values, the ranking observed for apparent density was inverted. The pore size distribution analysis showed that the dried tuna is highly heterogeneous in this regard, as expected for biological materials. The pore size distribution curve was so rugged for the air-dried sample that it was not possible to establish the peaks. However, it was possible to observe that most of the pores presented less than 15 μm. The freeze-dried product presented three major peaks of pores: at 5.5, 13 and 30 μm, the latter being the higher. The vacuum-dried product presented major peaks at 3.8, 4.1 and 4.8 μm. The pore size distribution curves confirmed the results obtained for apparent density, i.e., the freeze-dried product presents larger pores. After 6 months of storage, color and peroxide value of the dried product were measured. The freeze-dried sample presented the highest L* value, suggesting a lower darkening when compared to the other products. On the other hand, the freeze-dried fish presented the highest peroxide value, indicating highest oxidation of fats. Such result was related to the highest porosity of freeze-dried product, thus allowing a greater amount of oxygen to enter. Summarizing, freeze-dried tuna fish was the most porous among the obtained products, a characteristic that is associated with good rehydration capacity. In addition, freeze-dried tuna presented the best color. Nevertheless, it probably presents the shortest shelf-life due to high rate of lipid oxidation.

Cui et al. [6] compared the effect of freeze-drying, hot air-drying and microwave-vacuum/air drying on quality indices of garlic slices. Hot-air drying was performed at 60–65 °C and air velocity of 0.3 m/s. Freeze-drying was performed at a heating plate temperature of 45 °C and 5 mbar of pressure. The new method proposed in the study consisted in combining microwave-vacuum drying with air drying. The main part of the drying process was performed by microwave-vacuum technology with an actual measured power output of 400 W, pressure of 25 mbar and rotational speed of the turntable of 5 rpm. The microwave power level/time combination was: 100 %/7 min, followed by 50 %/8 min and finally 18 %/20 min. This step led the garlic slices to a moisture content of 10 % w.b. Then, to finish the drying, air at 45 °C was used to reduce the moisture content of the garlic slices until 5 % w.b. Since the characteristic garlic pungency is appreciated by consumers, the authors used a technique based on the measurement of the pyruvate content for expressing pungency. In addition, other quality characteristics viz. color, texture and rehydration ratio were analyzed. Results showed that freeze-dried product presented as much pungency as fresh garlic. Hot-air drying resulted in severe loss of pungency. The microwave-vacuum drying combined with air drying caused small loss of pungency (11.18 %), a value not as low as did freeze-drying (2.22 %), but much lower than hot air-drying (45.84 %). Regarding color, once again the freeze-dried product was better preserved than the other products, as expressed by higher lightness, lower redness and lower yellowness. Texture as represented by cutting force (g) was tougher for the hot air-dried product, followed by microwave-vacuum/hot air dried and freeze-dried. A tough texture is related to severe shrinkage and structural collapse, while a soft texture is associated with intact structure after drying. With concern to rehydration capacity, the three drying techniques yielded similar products. Such result was attributed to the low thickness (2–3 mm) of the garlic slices which allowed water to move freely through them. Summarizing, the microwave-vacuum-drying associated with hot air-drying showed to be a promising technique for obtaining high quality dried garlic at a lower cost when compared to freeze-drying.

Cui et al. [7] compared the effect of several drying techniques on pigment retention of carrots and Chinese chive leaves. Carrots were evaluated for their carotene content and chives for their chlorophylls' content. The drying processes were as follows: hot-air drying (air velocity of 0.3–0.5 m/s; 60–65 °C); freeze-drying (heating plate temperature of 45 °C, 5 mbar); and microwave-vacuum drying (25 mbar; 5 rpm) under several power level/time combinations (80–400 W/15–30 min), either alone or followed by hot-air drying (air velocity of 0.3–0.5 m/s; 40–45 °C) or conventional vacuum drying (55–60 °C; vacuum not informed). In addition, the effect of blanching in water at 90 °C for 3 min on the carrots and chives' pigment retention was evaluated. Results showed that freeze-drying and the various types of microwave-vacuum drying were comparable regarding carotene retention (94.7–97.8 %) and chlorophylls retention (94.7–99.4 % for total chlorophyll, 93.8–100 % for chlorophyll a, and 97.0–98.3 % for chlorophyll b). On the other hand, hot-air drying presented an inferior performance of retention (70.8 % of carotene, 38.3 % of total chlorophyll, 37.0 % of chlorophyll a, and 41.9 % of chlorophyll b). Blanching remarkably improved the carotene retention of hot-air dried carrots from 70.8 to 85.5 %, while no significant effect was observed for the other drying techniques. Summarizing, microwave-vacuum drying is a process that preserves vegetables natural pigments while presenting a reasonable cost. Furthermore, it does not require the use of a blanching pretreatment.

Devahastin et al. [9] compared the effect of different conditions of low pressure superheated steam drying (LPSSD) and vacuum drying on typical process and quality parameters of carrot cubes. The LPSSD process had been used previously for drying other materials and in that study it was tentatively used for food. The schematic diagram of a low-pressure superheated steam dryer is shown in Fig. 1. The absolute pressure inside the dryers was kept at 7, 10 or 13 kPa. The drying temperatures chosen were 60, 70 or 80 °C. It was observed that lower pressures and higher temperatures led to higher drying rates and lower drying times. In addition, vacuum drying removed moisture from the carrots faster than LPSSD did. The effect of temperature on the drying rates was greater than the effect of pressure, in the case of LPSSD. In the case of vacuum drying this behavior was not clear. When compared to previous studies, it can be affirmed that LPSSD and vacuum drying lead to higher drying rates than hot air drying and freeze drying. With regard to the quality parameters evaluated, it was seen that the use of higher pressures led to higher shrinkage. This finding shows that lowering the pressure during subatmospheric drying helps preventing the structural collapse of foods. Even though similar values of shrinkage were observed for the two techniques tested, the vacuum dried carrots shrank less uniformly than the LPSSD dried ones.

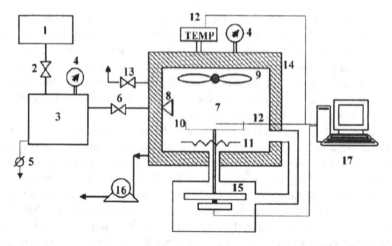

Fig. 1 Schematic of a low-pressure superheated steam dryer: *1* boiler; *2* steam valve; *3* steam reservoir; *4* pressure gauge; *5* steam trap; *6* steam regulator; *7* drying chamber; *8* steam inlet and distributor; *9* electric fan; *10* sample holder; *11* electric heater; *12* on-line temperature sensor and logger; *13* vacuum break-up valve; *14* insulator; *15* on-line weight indicator and logger; *16* vacuum pump; *17* PC with installed data acquisition card [9]

In addition, carrot that underwent LPSSD presented better rehydration capability than the vacuum dried carrot. This behavior could be explained by formation of a dense external layer during vacuum drying, which prevented the re-adsorption of water. With regard to the color of the final product, the LPSSD dried carrots were redder and lighter than those dried by vacuum drying. In conclusion, conventional vacuum drying leads to higher drying rates and shorter drying times than LPSSD. On the other hand, the final product was better for the LPSSD dried carrots.

Sunjka et al. [35] compared the effect of microwave-vacuum (MW-vacuum), microwave-convective (MW-convective) and ordinary convective drying on quality parameters of cranberries. The latter was used to generate a kind of control sample, being carried out with air temperature of 62 °C and air velocity of 1.0 m/s. MW-vacuum drying was performed at 3.4 kPa, power densities of 1.00 or 1.25 W/g of product and microwave modes of 30 s on/30 s off or 30 s on/45 s off. The sample layer comprised 3 cm. MW-convective drying was performed with air temperature of 62 ± 2 °C, air velocity of 1.0 ± 0.1 m/s, power densities of 1.00 or 1.25 W/g of product and microwave modes of 30 s on/30 s off or 30 s on/60 s off. The cranberries were previously cut into halves (MW-convective) or quarters (MW-vacuum) and osmotically dehydrated in high fructose corn syrup (76 °Brix) for 24 h at room temperature. Samples were dried to an approximate 15 % moisture content w.b. The quality parameters evaluated were: color (L*, a*, b*, C*, h* and ΔE); texture (toughness and Young's modulus) and sensory scores (color, texture, taste and overall acceptability). In addition, the microwave processes were evaluated for drying performance (kg of evaporated water/J of supplied energy) and cumulative energy efficiency (see [12] for details). Those authors observed that the MW-vacuum drying preserved in a better way the cranberries color, as denoted by higher L*, a*, b* and C* values and lower h* and ΔE values. Among the experimental conditions tested, the MW-vacuum drying performed at 1.00 W/g power density and 30 s/45 s on/off period provided products of best color. Regarding texture, once again the MW-vacuum drying was superior, providing chewier products (low toughness and low Young's modulus). Product temperature during MW-vacuum drying rose above 27 °C, which is the water evaporation temperature at 3.4 kPa. This was justified by the fact that the microwave power used was too high, evidencing the need for a more precise control of process conditions. Sensory quality was similar for all three drying techniques. Nevertheless, sensory scores were slightly superior in cranberries obtained by ordinary convective drying, probably due to the general nonuniformity observed in microwave dried cranberries. Finally, MW-vacuum drying proved to be more interesting from an energetic viewpoint than MW-convective drying, i.e., higher values of drying performance and cumulative energy efficiency. In this sense, the combination of vacuum and microwave has shown to be promising regarding color and texture, and energy aspects, while sensory quality of the generated products needed to be improved through optimization of the process conditions.

Yanyang et al. [45] compared the use of ordinary hot air drying to the use of hot air drying combined with microwave-vacuum drying for removing moisture from wild cabbage (*Brassica oleracea*). The quality of the final products was compared. Hot air drying was carried out at 60–85 °C with air velocity of 1.15 m/s. Microwave vacuum drying was performed at input power ranging from 1,400 to 3,800 W and pressure ranging from 2.0 to 2.5 kPa. Final product quality was expressed as sensory characteristics, retention of chlorophyll and ascorbic acid, color and microbial index. Sensory tests showed that the most suitable process conditions were to use hot air drying until the product reached 15 % of moisture w.b. and then use microwave vacuum drying at 1,900 W and 2.0 kPa. It was observed that drying time decreased drastically (~31 %) when compared ordinary hot air drying to hot air drying associated with microwave vacuum drying. In addition, chlorophyll and ascorbic acid retention was improved when using the hybrid method. Color as expressed in the CIE L*a*b* color space was better for the hot air/microwave vacuum dried wild cabbage, i.e., higher greenness. Finally, the microbiological quality of both final products was satisfactory. Summarizing, it was considered advisable to complement the traditional hot air drying with microwave vacuum drying for reducing drying time and improving dried wild cabbage quality.

Caro-Corrales et al. [5] analyzed how varying the conditions of controlled low-temperature vacuum dehydration (CLTVD) and tunnel drying affected selected quality and engineering aspects of mashed potato slabs. On a second moment, both techniques were used under optimized conditions to dry the mashed potatoes in a comparative way. The CLTVD technique was developed by King et al. [17]. Here, it was performed in a freeze-dryer, which was operated above water freezing point for the equivalent pressure. In practical terms, temperatures ranged from −1 to −9 °C and pressure from 0.67 to 1.33 kPa. The tunnel drying was performed at 40–60 °C and air velocity of 2 m/s. Product thickness varied form 0.26 to 0.78 mm. The potatoes were peeled, cubed and canned (121.1 °C/15 min) previous to mashing. The quality features analyzed were total color difference (ΔE), water adsorption (rehydration) capacity and bulk density. The engineering features comprised critical moisture content, drying rate, mass transfer coefficient and effective diffusivity. It was observed that lower product thickness (0.26 mm) and higher drying temperature (60 °C) led to faster tunnel drying. In addition, lower product thickness (0.26 mm) and lower chamber pressure (0.67 kPa) led to faster CLTVD drying. The critical moisture content is the amount of water contained in the product at the transition from the constant rate drying period to the falling rate drying period. This feature is useful for determining the beginning of the falling rate period, since during this period the drying rate decreases progressively. The critical moisture content during tunnel drying decreased with increase in temperature. During CLTVD, it decreased with increase in pressure and thickness. The drying rate was affected by temperature during tunnel drying and by pressure during CLTVD. The mass transfer coefficient was not affected by different process conditions. At all process conditions, drying rates were higher (one order of magnitude) and mass transfer coefficients were much higher (four orders of magnitude) for tunnel drying when compared to CLTVD. On the other hand, when it comes to product quality, CLTVD was superior, i.e. lower bulk density, higher water adsorption and lower ΔE. When the optimum conditions were tested, the results were repeated, i.e., tunnel drying was better in terms of engineering while CLTVD provided better product quality. Thus, it can be concluded that CLTVD represents an interesting option for drying heat-sensitive foods.

Regier et al. (2005) compared the effects of hot air drying, freeze-drying and microwave-vacuum drying on total and selected carotenoid content of *Nutri Red* carrots. This variety of carrots presents enhanced lycopene content. In addition, the stability of carotenoids to long term storage was assessed. In the lab, two sets of experiments were performed: (1) the carrots were peeled, sliced, blanched at 95° for 65 s, frozen, thawed and dried; (2) the carrots were peeled, cut into 1, 3 and 6 mm height, blanched at 90 °C for 2 min and dried. The material for all the analyses, except from the analysis of the effect of slice thickness on carotenoid retention, was generated by the first set of experiments. Hot air drying was performed in two dryers: a laboratory dryer and a pilot scale dryer. Both were operated with recirculating air and the laboratory dryer was also operated with inert nitrogen for comparison purposes. In the former, temperatures of 60 and 70 °C and gas velocities of 2.5 m/s were used. In the latter, temperature ranged from 50 to 90 °C and air velocity of 4 m/s. The relative humidity of the air was always 8 ± 2 %. The process was interrupted at 0.1 g/g dry matter product moisture content. Freeze-drying was conducted at 6 Pa, condenser temperature of −60 °C and heating plate temperature of 30 °C. Microwave-vacuum drying was carried out at 5 kPa and two combinations of power input: a continuous process at 400 W or a combined process beginning at 600 W for 75 min followed by the rest of the process at 240 W. For studying the effect of storage on dried carrots carotenoid content, they were stored at 25 °C protected from light and surrounded by air or nitrogen. It was shown that blanching barely affected the carotenoid content of fresh carrots. With regard to hot air dried carrots, lycopene was preserved until 90 °C. On the other hand, β-carotene was destructed above 70 °C. The use of nitrogen inside the drying chamber did not show any protective effect on carotenoids when compared to air. Hot air drying improved the carotenoid extractability, being this effect more pronounced for thinner slices and lower drying times. The freeze-dried product retained the carotenoids completely. The carrots dried by microwave-vacuum at 400 W of power not only had all carotenoids preserved, but their content was enhanced due to improved extractability. On the other hand, the combined process (600 W/240 W) reduced the carotenoid content by 20 %. It is needed to say that the microwave-vacuum drying was much shorter than the other processes. Finally, the carrot slices dried with hot air at 50 and 60 °C and then stored under nitrogen had their carotenoid preserved, while those surrounded by air lost about 20 % of their carotenoids. Summarizing, all processes were equivalent regarding carotenoid retention when mild temperature were used. However, microwave-vacuum drying offers higher drying rates than the other techniques.

Rodríguez et al. [32] compared freeze-drying with microwave-vacuum drying in terms of engineering and quality aspects during the drying of mushrooms. Dried product quality was expressed as water sorption capability and micrographic images. The evaluated engineering aspects were drying kinetics and effective diffusivity. Freeze-drying was performed at a heating plate temperature of 50 °C, an initial product temperature of -20 °C and a pressure varying from 0.2 to 30 mmHg. Microwave-vacuum drying was performed in two major fashions: with and without product temperature control. The former was performed at 30 mmHg of pressure, product temperature limited to 100 °C and power level ranging from 60 to 240 W. The latter was performed at the same pressure and power conditions used in the uncontrolled product temperature process, but here product temperature was limited to 25, 30, 40 or 50 °C. In fact, all "microwave processes" were divided in two phases: I, without applying microwaves and at very low pressure (0.2 mmHg), used for the first 30 min; and II, applying microwave at the pre-set levels and at a pressure of 30 mmHg until constant weight was attained. According to the results obtained, the freeze-drying rates were higher for lower operation pressures. Regarding the microwave processes, product temperature control was shown to affect the drying kinetics in the following fashion: initially, the controlled temperature process presented lower drying rates compared to the uncontrolled temperature process, while at the final stages of drying this behavior was inverted. Furthermore, the product that had its temperature controlled reached lower moisture content at the end of the process. With regard to the effective diffusivity value, the higher the microwave power level used, the higher the effective diffusivity. A clear increase in effective diffusivity was observed from phase I (without microwave) to phase II (with microwave) of drying. In parallel, the lower the pressure during freeze-drying, the higher the effective diffusivity. Values of effective diffusivity ranged from 3.6×10^{-10} to 2.9×10^{-9} m^2 s^{-1} for the microwave process without product temperature control; 1.15×10^{-10} to 9.30×10^{-10} m^2 s^{-1} for the microwave process with product temperature control; and 1.05×10^{-10} to 1.60×10^{-10} m^2 s^{-1} for the freeze-drying process.

Concerning product quality, it was observed that lower operation pressure during freeze-drying led to higher dried product sorption capacity at low relative humidity (<40 %). For the microwave dried product, the sample dried under controlled temperature presented worse rehydratability than that obtained without temperature control. Scanning Electron Microscopy images of fresh, freeze-dried and microwave-vacuum dried mushrooms were similar, all with negligible shrinkage. Summarizing, microwave-vacuum drying yielded high drying rates, especially when product temperature was controlled, besides providing the product with a quality similar to that of freeze-dried mushrooms.

Suvarnakuta et al. [36] compared the effect of LPSSD, vacuum drying and hot air drying on the drying kinetics and β-carotene content of carrots. Product temperature was monitored throughout the drying processes. Correlations between product moisture content, temperature and β-carotene content were tentatively established by using empirical models. In the lab, the carrots were peeled and diced into 1 cm^3 cubes. The samples consisted of a single layer of carrot cubes. LPSSD was carried out at a steam pressure of 7 kPa and steam temperatures ranging from 60 to 80 °C. The flow rate of steam was roughly 26 kg/h. Vacuum drying was carried out at the same pressure and temperature conditions used for LPSSD. Hot air drying was performed at the same temperatures mentioned above, although at atmospheric pressure and with the air being blown inward the chamber at 0.8 m/s. As can be inferred from the results, higher operation temperature yielded higher drying rates. Product temperature increased during drying, especially at the final stages when most of the water has been evaporated and the heat received from the equipment is almost exclusively used for heating the material. Comparing the drying times for the three processes, vacuum drying took shorter than the other two. The time difference between LPSSD and vacuum drying tends to decrease with increasing temperature. Comparing hot air drying and LPSSD, one concludes that the drying times were roughly the same. Nevertheless, hot air drying was faster at initial stages of drying and LPSSD was faster at the final stages, when a more porous structure allowed freer water diffusion. When it comes to β-carotene retention, the best technique was LPSSD, followed by vacuum drying and hot air drying. Good correlations ($R^2 \geq 0.90$) were established between carrots β-carotene content and temperature/moisture content. On the basis of the magnitude of the equations' terms, it is possible to affirm that temperature affected more the β-carotene content than the moisture content did. As a conclusion, LPSSD was more effective in preserving the β-carotene content of the carrots, albeit the associated drying times were higher than for vacuum drying.

Hawlader et al. [13] studied the drying of guava and papaya by heat pump drying (HPD), vacuum-drying and freeze-drying. Engineering and quality aspects were evaluated, being the former only for the HPD process. Guava and papaya cubes measuring 1 cm^3 were dried. HPD was performed either in air, nitrogen or carbon dioxide. A lab-scale heat pump dryer was operated at relative humidity of 10 %, temperature of 45 °C and air velocity of 0.7 m/s for 8 h. Freeze-drying was performed at 10 °C and 4.6 mmHg during 24 h. Vacuum drying was performed at 45 °C and 0.15 bar during 24 h. Dried fruit quality was expressed in terms of color, porosity, rehydration capability, texture and vitamin C content. Results showed that the use of inert gases, such as nitrogen and carbon dioxide, increases the drying rate during HPD of both fruits, as confirmed by the high values of drying constant obtained by fitting the kinetic data to the Page model. Alternatively, the Fick's second law of diffusion for a flat plate was used to fit the kinetic data. As expected, effective diffusivities repeated the behavior observed for drying constants for both fruits, i.e., the highest values were obtained for nitrogen HPD, followed by carbon dioxide HPD and air HPD. As mentioned before, when quality aspects were taken into account, the other two drying techniques were considered. Color, as expressed by total color difference (ΔE), was better preserved by freeze-drying, in the case of guava, and by nitrogen HPD in the case of papaya. With regard to bulk density, the lower its value, the better the dried product quality. The lowest values of bulk density were observed for freeze-dried fruits. The measures of porosity were expected to behave in the opposite fashion, which was confirmed in the case of guava. On the other hand, freeze-dried papaya presented the lowest porosity, probably due to limitations of the porosity measurement technique. Vacuum dried papaya presented the highest porosity. HPD in inert gas yielded products of intermediate bulk density and porosity, while HPD in air yielded the worst quality products in this regard. Scanning electron microscopy images showed that freeze-dried fruits presented the most porous structure. Vacuum dried and inert gas heat pump dried fruits presented a pore structure similar to each other, while air heat pump dried fruits presented only a few pores. In terms of rehydration capacity, the products presenting higher porosity were the ones that better rehydrated, except from the vacuum dried products. The results of the firmness analysis showed considerable variation. One of the few evidences was the high toughness of the vacuum dried product, probably due to a case hardening effect, which might explain its low rehydration capacity. Finally, vitamin C retention was higher for the freeze-dried product, followed by the vacuum dried, the carbon dioxide heat pump dried, the nitrogen heat pump dried and the air heat pump dried. In sum, the techniques using vacuum still present advantages, but new techniques like HPD are promising in generating high quality fruits at a relatively low cost, especially when operated under inert gases.

Qing-guo et al. [29] compared the effect of several drying techniques on quality aspects of granular edamame (*Glycine max* (L.) Merrill). The edamames were pretreated by immersion in a 3 % salt solution for 1 h. Then, they were drained and arranged in a single layer comprising 1.0 kg and dried either by: hot air drying at 1.5 m/s air velocity, 20 % air relative humidity and 70 °C air temperature; microwave-vacuum drying at a power density of 3.5 kW and 6.325 kPa of absolute pressure; freeze-drying at a condenser temperature of −20 to −45 °C, a heating plate temperature of 60 °C and 0.05–0.27 kPa of absolute pressure; and hot air drying at 70 °C and 1.4 m/s air velocity for 20 min followed by microwave-vacuum drying at a power density of 2.8 kW and 6.325 kPa of absolute pressure. Quality was expressed as water, vitamin C and chlorophyll contents, surface color, shrinkage, rehydration capacity, texture and microstructure. Results showed that microwave-vacuum drying presented the highest drying rate, followed by hot air drying/microwave-vacuum drying, hot air drying and freeze-drying. Microwave-vacuum trials took less than 1 h to dry 1 kg of edamames. Regarding nutrient retention, freeze-drying was shown to preserve in a better way the edamames' vitamin C and chlorophyll. Even though the microwave-vacuum trials did not preserve as much vitamin C and chlorophyll as freeze-drying, they were much better than hot air drying, preserving roughly twice as much of both nutrients. With regard to the changes in edamame color during drying, the samples became darker, greener and yellower, except from the freeze-dried samples. In time: the authors attributed the color changes to the Maillard reaction, which does not take place at the temperatures used for freeze-drying. Air dried samples suffered the severest color changes. With concern to shrinkage, once again the best quality, i.e., lowest shrinkage, was observed for freeze-dried edamames, while the highest shrinkage was observed for hot air dried edamames. The best rehydration capacity was obtained by the freeze-dried product, and the worse for the hot air dried one. The rehydration process was accelerated by augment in temperature. Texture tests showed that the best crispness was attributed to the freeze-dried edamame, while the best chewiness was attributed to hot air/microwave-vacuum dried edamame. In general terms, it was concluded that freeze-drying provides proper preservation of nutrients and color in the final product, but leads to poor chewiness and is associated with high costs. Hot air drying is a cheap method, but promotes nutrients' destruction and color degradation. On the other hand, microwave-vacuum drying is relatively low cost and provides the edamames with a good quality after drying, especially when preceded by hot air drying. When both techniques are combined, a product of great chewiness is generated.

Giri and Prasad [11] compared the effect of microwave-vacuum drying (MVD) and hot air drying on the drying kinetics, rehydration properties and microstructure of mushrooms. In addition, the MVD was optimized as a function of the drying constants of two selected models. MVD was performed at variable microwave power (115–285 W), pressure (6.5–23.5 kPa) and samples thickness (5.8–14.2 mm). Hot air drying was performed at 50, 60 or 70 °C and air velocity of 1.5 m/s. Drying curves showed that MVD was much faster than air drying. For the former, results indicated that higher microwave levels yielded faster drying. For the latter, it was observed that higher air temperatures yielded faster drying. Among the studied variables, microwave power level markedly affected the drying times, while sample thickness and especially system pressure did not present much effect. The plot of drying rate versus moisture content showed that MVD did not present a constant rate period, in opposition to hot air drying. Considering the models used for fitting the drying curve, the Page model (Eq. 10 in chapter Studies on Conventional Vacuum Drying of Foods) presented a better fit than the exponential model, i.e., higher values of coefficient of determination. The values of drying constant (min^{-1}) of the two models increased with increasing temperature for hot air drying and with increasing power level/decreasing pressure for MVD. Analysis of variance showed that polynomial models provided a proper explanation of the effect of the studied variables on drying constants and rehydration capacity, although significant lack of fit was observed. Drying constants of the Page and exponential models were both mostly affected by power level and thickness.

Response surface plots showed that higher drying constants were obtained for higher power level and lower sample thickness. On the other hand, rehydration was mostly affected by system pressure and thickness, with higher vacuum and thinner slices giving better rehydration. Furthermore, rehydration curves showed that the mushrooms that rehydrated better were those dried by MVD at high power levels and high vacuum. Finally, micrographs confirmed that mushrooms dried by MVD present a more preserved and porous structure than those dried by hot air, especially under higher vacuum. In conclusion, MVD showed suitable for yielding dried mushrooms in a feasible drying time and with good rehydration properties, especially when high power level, low system pressure and low sample thickness were used in combination.

Li et al. [20] compared the effect of freeze-drying and microwave-vacuum drying/vacuum drying (MVD/VD) on the allicin content, color, microstructure and other properties of garlic slices. Allicin is a functional compound that is generated by the catalyzing action of the alliinase enzyme on alliin when the garlic tissue is disrupted. In addition, the garlic slices prepared by MVD/VD were microencapsulated and submitted to in vitro digestion. The objective here was to evaluate if the microencapsulation was able to protect the alliinase enzyme through the digestion in the stomach. Freeze-drying was carried out at a heating plate temperature of 45 °C and 10 Pa of absolute pressure. MVD was performed at power levels ranging from 94.0 to 376.1 W and time ranging from 3 to 15 min, until the product moisture content achieved 15 %. Preliminary tests where MVD pressure was varied showed that above 6 kPa the temperature of water boiling is above that tolerated by alliinase. Then, MVD pressure was fixed at 6 kPa. The MVD was followed by VD at 40 °C and 4 kPa until the product moisture content achieved less than 5 %. The goal was to use VD at the end of the drying process in order to avoid product overheating and consequent alliinase destruction. Response surface methodology was used to optimize the MVD as function of thiosulfinates content, from which allicin is the major compound. The optimum MVD conditions as regards time/power level were: 3 min/376.1 W followed by 3 min/282.1 W, then 9 min/188.0 W and finally 3 min/94.0 W. Such conditions yielded a thiosulfinates retention of 90.2 %. Ulterior VD was not considered to degrade any thiosulfinates. When compared to freeze-drying, MVD/VD generated dried garlic of roughly similar color and thiosulfinates retention, with the advantage of presenting much higher drying rates. Micrographs showed that the microencapsulation was successful, generating continuous and smooth garlic power granules. Thiosulfinates measurement after exposure of garlic powder to gastric juice confirmed that the microencapsulation process was able to protect alliinase (thiosulfinates retention > 98 %). The release of thiosulfinates in a simulated intestine medium obeyed a zero-order kinetic followed two first-order kinetics. Concluding, that study showed that it is possible to obtain garlic powder with high functional compounds content by using MVD/VD. The final product of MVD/VD presented quality similar to that of a freeze-dried product.

Panyawong and Devahastin [27] elucidated the drying kinetics and studied the degree and the pattern of carrot cubes shrinkage submitted to LPSSD and vacuum drying under various temperatures. Both drying methods were performed at 60, 70 and 80 °C and pressure of 7 kPa until the sample achieved a moisture content of 0.07 kg/kg (d.b.). The degree of shrinkage was expressed as dimensionless volume, i.e., the ratio of sample volume at any time to initial sample volume. The pattern of shrinkage was expressed as Heywood shape factor [44] that is the ratio of volume to the equivalent projected-area diameter of the cube. Results confirmed that the higher the temperature, the higher the drying rates, as expected. LPSSD yielded lower drying rates than vacuum drying for all temperatures tested. Dimensionless volume was found to decrease with time, confirming that the sample volume decreases with time, i.e., the degree of shrinkage increases. For all techniques and drying temperatures the degree of shrinkage followed a second order nature. Comparing the two techniques in terms of changes in degree of shrinkage during drying, the authors concluded that: on initial stages of drying, shrinkage of carrot undergoing LPSSD was less than that of vacuum drying, while at the end of drying this behavior inverted, probably due to case hardening at the end of vacuum drying. The degree of shrinkage at 60 °C was significantly different from that at 70 or 80 °C for both LPSSD and vacuum drying. The degree of shrinkage did not vary for different drying techniques, though. With regard to the pattern of shrinkage, it was observed that the Heywood shape factor value varied little on initial stages of drying and then decreased. The first stages, categorized as uniform deformation period, were longer for lower drying temperatures, suggesting that the use of lower drying temperatures leads to a better or longer preservation of the product shape. In addition, the uniform deformation period was longer for LPSSD than for vacuum drying, due to slower drying rates in the former. The slopes of the non-uniform deformation period, that were considered as being a rate of deformation, were higher for higher temperatures and also higher for vacuum drying compared to LPSSD. In conclusion, though the degree of shrinkage was not different for different drying techniques, the pattern of shrinkage was different in this sense. During vacuum drying, the carrots shrank faster and less uniformly than during LPSSD.

Thomkapanich et al. [37] studied the effect of turning off periodically (intermittency) the heat or the vacuum source during low pressure superheated steam drying (LPSSD) and vacuum drying of banana chips. Process features and final product quality were analyzed after heat or vacuum intermittency. Regarding the continuous processes, those authors found that LPSSD took longer than vacuum drying, especially at lower process temperatures. During intermittent-heat LPSSD, slightly higher drying rates were observed when compared to the continuous LPSSD. This effect was more pronounced at higher temperatures (90 °C). On the other hand, in the case of intermittent-heat vacuum drying, the drying rates were lower than those observed during the continuous vacuum-drying. For the intermittent-vacuum processes, both of the techniques studied presented higher drying rates when compared to the continuous processes. Such behavior was attributed to the formation of a network of interconnected pores by the decompression of pressure that takes place during intermittent-vacuum drying, allowing freer moisture diffusion. Regarding energy savings, the intermittent-heat LPSSD saved more energy than the intermittent-heat vacuum drying for all temperatures studied. In the case of the intermittent-vacuum process, though, it was observed that the energy savings in the case of LPSSD were similar to those observed during vacuum drying. The quality of dried banana chips was evaluated on the basis of the following parameters: color, texture, shrinkage and ascorbic acid content. It was observed that the original color of the bananas was better preserved by using intermittent-heat LPSSD. On the other hand, the use of intermittent vacuum affected the two methods equally: the entrance of oxygen in the drying chambers during the off period led to enzymatic browning, and ultimately to lower lightness in the samples. The texture of the dried bananas was not significantly affected by different methods or temperatures, while the use of intermittent vacuum led to higher hardness. Similarly, shrinkage was not affected by different methods, while the use of intermittent vacuum led to higher shrinkage. Finally, ascorbic acid retention in the bananas dried by both techniques was higher under intermittent-heat and lower under intermittent-vacuum. In conclusion, the use of heat and vacuum intermittency during LPSSD and vacuum drying led to various results. Markedly, this new approach leads to energy savings and in some cases, better product quality.

Yaghmaee and Durance [42] evaluated the decontaminating effect of microwave-vacuum drying alone or combined with atmospheric microwave drying on carrot and parsley. The drying kinetics was also generated. Microwave-vacuum drying was conducted at 8–15 kPa and power level/drying time of 1.8 kW/58 min for carrot or 1.5 kW/20 min for parsley. The hybrid method consisted in starting the drying (of carrots only) at atmospheric pressure and 1.8 kW microwave power for 12 min followed by making vacuum in the chamber at 8–15 kPa for 45–48 min with the same microwave power. The products container, a plastic cylinder, was turned at 15 rpm on its axial axis. The samples had their surface temperature measured during drying. The microbial counts performed were total aerobic count, molds and yeasts. The drying kinetics of both techniques was roughly the same. As expected, higher product temperatures promoted higher microbial reduction. In this sense, the atmospheric microwave process heated more the product and consequently killed more microorganisms. At the end of the atmospheric microwave process the product temperature reached 85 °C. The exception was the molds count, which decreased more during microwave-vacuum drying probably due to different initial microbial load and/or variation in mold species. The microbial counts decreased steadily with the advance of drying in the case of microwave-vacuum process. On the other hand, the atmospheric microwave process promoted a sharp decrease in microbial count. In general terms, the microbial load of the microwave atmospheric/microwave-vacuum dried products was significantly lower than that of the microwave-vacuum dried product. This finding led to the conclusion that starting the microwave-vacuum drying process at atmospheric pressure for a few minutes works as a pasteurization step, assuring more safety to the final product.

Acevedo et al. [1] compared the effect of various vacuum drying processes on the microstructure of apple slices and their relation to selected quality aspects. Apple discs were dried by: vacuum drying at 50 °C and 25 inHg of absolute pressure, alone (VD) or associated with previous blanching for 3 min in live steam (B/VD); freeze-drying at a condenser temperature of -70 °C and 6×10^{-3} bar of absolute pressure following slow freezing (SF/FD) or fast freezing (FF/FD). Sorption capacity, microstructure, color, texture and image analyses were performed in the product right after drying and also after storing at 70 °C protected from water loss, after being equilibrated to several levels of relative humidity. In addition, powdered apple was used for comparison purposes. The sorption capacity of the samples was comparable, except from the FF/FD samples that presented increased sorption especially above 33 % of relative humidity. With regard to structural changes during drying, vacuum-dried apples suffered a severer cellular collapse as compared to freeze-dried apples. Among freeze-dried samples, the FF/FD sample presented a more preserved structure than the SF/FD sample. After storage at 70 °C for 36 h, all products presented further structural modifications. Once again, the degree of structural modification was higher for vacuum-dried samples. Concerning apple color, more significant color changes were observed in the vacuum-dried apples, an effect which was attenuated by blanching. Freeze-dried samples were more preserved in terms of color, especially the FF/FD apples. Increasing relative humidity and increasing time of storage led to increased color changes. Texture analysis showed that hardness decreased with increasing relative humidity. The hardest apples were the vacuum-dried ones, due to increased shrinkage. Summarizing, all drying techniques affected the microstructure of the apples. The changes in microstructure impacted the quality of the final product, being vacuum-drying inferior to freeze-drying in this sense. Blanching was able to improve the vacuum-dried product quality in some regards. Finally, fast freezing was better than slow freezing for the quality of the freeze-dried product.

Deng and Zhao [8] compared the effect of innovative osmotic pretreatments combined with hot air drying and freeze-drying on the quality of apples. The osmotic pretreatments consisted in immersion in high-fructose corn syrup and a mixture of calcium lactate and gluconate combined with: agitation, ultrasound or pulsed vacuum. The latter consists in placing the system in a vacuum chamber and applying vacuum for some minutes repeatedly. Then, the apples were dried by hot air at 55 °C, 3.86 m/s air velocity and 20 % air relative humidity; and freeze-drying at an initial sample temperature of about −45 °C and 13.3 Pa of absolute pressure, in order to dry the apples close the their glass transition temperature (−45 °C). The final products were evaluated for their glass transition temperature, texture, shrinkage, rehydration, color, microstructure, calcium absorption profile, water activity and moisture content. Results showed that pretreatment by ultrasound yielded lower values of moisture content/water activity and higher values of glass transition temperature. Pulsed vacuum promoted the same effect, but with lower intensity. Since glass transition temperature reflects product stability, higher values of this parameter are associated with higher shelf life. In this sense, hot air dried apples seem to be more stable than freeze-dried apples. With regard to texture, air drying yielded products of higher hardness and crispness. Nevertheless, the air dried product shrunk more and rehydrated worse than the freeze-dried product. Ultrasound led to higher hardness in the freeze-dried product and pulsed vacuum led to higher shrinkage in the air dried product. Concerning microstructure, freeze-dried apples presented more pores than hot air dried apples, though both showed structural collapse. Different pretreatments exerted little effect on color, while drying methods markedly affected color with air drying promoting more browning than freeze-drying. Nevertheless, both dried products were lighter than untreated apples. Calcium was found to be well distributed throughout the samples, which means that calcium contained in the osmotic solution impregnated into the apples. In conclusion, the authors reported that the osmotic pretreatment combined with ultrasound and pulsed vacuum prior to drying yields apples of better quality in some cases, but further studies needed to be conducted in order to optimize pretreatment conditions. Once again, freeze-drying yielded a product of better quality than air drying, except from product crispness and shelf life.

Hiranvarachat et al. [14] studied the drying kinetics, the changes in product temperature and β-carotene content, and the formation of 13-*cis*-β-carotene during hot air drying, vacuum drying and low pressure superheated steam drying (LPSSD) of carrots. The main objectives were to investigate the degree of β-carotene thermal degradation and the conversion of the naturally occurring all-*trans*-β-carotene into 13-*cis*-β-carotene by isomerization during drying, this one associated with reduced antioxidant activity. Product antioxidant capacity was measured during drying by the TEAC procedure. All drying methods were performed at 60, 70 and 80 °C. Pressure used was either 7 kPa in the case of vacuum drying and LPSSD or the atmospheric pressure in the case of hot air drying. Air velocity was fixed at 0.8 m/s for the latter. Carrot cubes were dried to a moisture content of 0.1 kg/kg (d.b.). Results showed that the higher the drying temperatures, the higher the drying rates, as expected. Vacuum drying yielded higher drying rates than hot air drying and LPSSD, although such difference was more subtle at higher drying temperatures. Product temperature increased from the ambient temperature to the temperature established for each process. The product temperature curves were useful for correlating to β-carotene retention. Three moments markedly affected the β-carotene retention: when carrot reached the temperature of lipoxygenase activation (45–60 °C), in the case of hot air drying, because this enzyme catalyzes the aerobic oxidation of β-carotene; above 60 °C, in the case of vacuum drying, due to thermal degradation of β-carotene; and after 120 min of drying, in the case of LPSSD, when a significant product temperature rise was observed, especially at the drying temperatures of 70 and 80 °C, promoting thermal degradation in the product. With regard to the formation of 13-*cis*-β-carotene, it was observed that changes in drying temperature or drying method did not present much effect. *Cis/trans* β-carotene proportion ranged roughly from 0.01 to 0.1. Regarding product antioxidant activity, it was observed a continuous decrease in this parameter with the advance of hot air drying, being this effect more pronounced at higher drying temperatures. In the case of vacuum drying and LPSSD, a drop in the antioxidant activity was observed when product moisture content reached 1 kg/kg and 2 kg/kg (d.b.), respectively. Comparing the three techniques, those using vacuum yielded final products of consistently higher antioxidant activity than hot air drying. In sum, thermal degradation of β-carotene was found to be more significant than isomerization degradation. In addition, LPSSD at 60 °C provided dried carrots with the highest β-carotene content and the highest antioxidant capacity among the drying techniques and conditions studied.

The drying of collard leaves was studied by Alibas [2] by conducting hot air drying, microwave drying and vacuum drying under controlled conditions. The effect of drying on moisture content, ascorbic acid content and color was evaluated. Hot air drying was performed at 1 m/s air velocity and temperature ranging from 50 to 175 °C. Microwave drying was performed at power level ranging from 350 to 1,000 W. Vacuum drying was conducted at pressure ranging from 0.4 to 100 mmHg and temperature of 50 or 75 °C. The collard leaves moisture content was reduced to 0.1 kg/kg (d.b.) in all cases. Results showed that microwave drying took from 2.5 to 7.5 min to dry 25 g of leaves. Higher power levels yielded lower drying times. Microwave drying rates decreased with the advance of drying as a result of water removal, since microwaves act directly on water molecules. Hot air drying took from 8 to 210 min to dry 25 g of leaves. Higher air temperatures yielded lower drying times. Hot air drying rates decreased with the advance of drying, since the residual bound water is more difficult to remove. Vacuum drying kinetics was affected by changes in process temperature and pressure. Drying times ranged between 35 min (75 °C/0.4 mmHg) and 195 min (50 °C/100 mmHg). Vacuum drying rates also decreased with increasing drying time, due to the same reason mentioned for hot air drying. Comparing the three techniques, microwave drying yielded the highest drying rates while hot air drying yielded the lowest ones. All drying kinetics were well fitted by the Page model (Eq. 10 in chapter Studies on Conventional Vacuum Drying of Foods), with higher drying constants (k) for microwave process, as expected. As regards ascorbic acid retention, the best results were provided by microwave drying at 750 W, followed by vacuum drying at 75 °C/0.4 mmHg and finally air drying at 50 °C. The results for color preservation as expressed in terms of difference from the color of fresh collard were similar to the results for ascorbic acid content. Summarizing, microwave drying at 750 W was considered the best process for obtaining high quality dried collard leaves in short drying time.

Rahman et al. [31] compared the effect of different drying methods and different drying conditions on the allicin potential of garlic. Allicin is a bioactive compound believed to present antimicrobial and anticancer activity, among others. Three processes were compared: air drying, vacuum drying at 80 kPa and drying in nitrogen atmosphere. It was observed that vacuum drying at 50 and 60 °C caused the lowest loss of allicin potential, similar to air drying at 50 °C and drying in nitrogen atmosphere at 40 °C. Regarding drying rate, vacuum drying led to intermediate values, i.e., lower than those obtained by air drying but higher than those obtained for drying in nitrogen atmosphere at the same temperature. With regard to effective moisture diffusivity, vacuum drying presented lower values than air drying when the drying temperature was 50 °C, but higher values for 60 and 90 °C.

Hossain et al. [15] studied the effect of three drying methods on the content of total and selected phenolics and antioxidant activity of six aromatic herbs. Air drying was performed at ambient temperature (~14 °C) in a dark and ventilated room for 3 weeks. Vacuum drying was carried out at 70 °C and 600 mbar of vacuum for 16 h. Freeze-drying was conducted at a condenser temperature of −54 °C and a pressure of 0.064 mbar during 72 h. All dried samples and also the fresh sample were freeze-stored at −20 °C inside vacuum pack pouches. The herbs comprised rosemary, oregano, marjoram, sage, basil and thyme. Air-dried samples presented the highest total phenolic content, followed by vacuum-dried, freeze-dried and fresh samples. For some herbs, total phenolic content of vacuum-dried and freeze-dried samples were statistically equal. Storage time exerted no effect on total phenolic content of the dried herbs. Fresh herbs, on the other hand, presented an increase in phenolics, which was attributed to chilling injury. The exception in this case was thyme, due to the fact that it presents a woody structure that does not suffer chilling injury. For antioxidant capacity as measured by the FRAP and ORAC methods, results roughly similar to those obtained in the total phenolic assessment were observed. The rosmarinic acid analysis showed that the highest amounts of this antioxidant compound were retained in air-dried samples, followed by vacuum-dried, freeze-dried and fresh ones. Good correlations were established between: total phenolics and antioxidant capacity as measured by the FRAP method (0.921); total phenolics and antioxidant capacity as measured by the ORAC method (0.672); antioxidant capacity as measured by the FRAP method and antioxidant capacity as measured by the ORAC method (0.806). Summarizing, drying of herbs improves the extractability of phenolic compounds with antioxidant properties. In this context, vacuum drying methods provided good quality products in relatively short time.

Purnama et al. [28] compared the effect of hot air drying, microwave-vacuum drying and freeze-drying on the quality of shredded ginseng roots. Hot air drying was carried out at 38 °C during 12 h. Microwave-vacuum drying was performed 0.8, 1.3 and 1.8 kW for 26, 12 and 10 min, respectively along with pressure of 43.7 mmHg. Freeze-drying was performed at a condenser temperature of −50 °C and a pressure of 0.2 mmHg during 4–7 days. Quality was expressed as final product moisture content, water activity, visual appearance, total porosity/pore size and content of total and selected ginsenosides. These are the ginseng bioactive compounds. Results indicated that the freeze-dried product presented the lowest moisture content and water activity, while the samples dried by the other methods presented roughly similar results for these parameters. In visual terms, hot air drying yielded a darkened and shrunken product, while freeze-drying and microwave-vacuum drying yielded products of more natural appearance. Total porosity was much higher for the freeze-dried product (77.15 %) than for the air dried one (31.08 %), while microwave-vacuum dried ginseng presented intermediate porosity (37.89–41.14 %). In addition, freeze-drying led to the formation of a greater number of large pores (>1.5 μm). With regard to the retention of bioactive compounds, air drying preserved significantly less ginsenosides than the other methods. In fact, microwave-vacuum drying at 1.8 kW promoted the highest retention of total ginsenosides. Selected ginsenosides were also assessed in the final product. Results showed that, generally speaking, individual ginsenosides content did not differ from one treatment to another. The exceptions were ginsenosides Rg1 and Rb1 with microwave-vacuum drying at 1.3 kW and hot air drying promoting

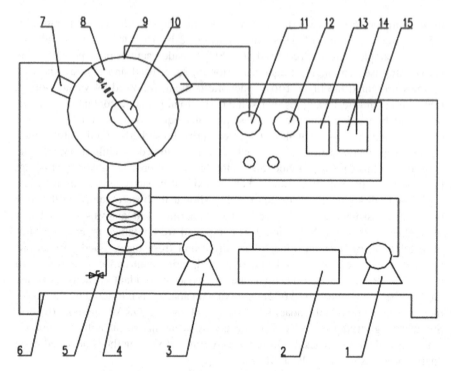

Fig. 2 Schematic of a microwave-assisted freeze-dryer: *1* refrigerator; *2* cooler; *3* vacuum pump; *4* cold trap; *5* blowdown valve; *6* equipment support; *7* magnetron; *8* drying cavity; *9* infrared thermoscope; *10* observation window; *11* thermometer; *12* pressure detector; *13* PLC controller; *14* power regulator; *15* control box [43]

the highest destruction of these compounds, respectively. In sum, microwave-vacuum drying at high power level (1.8 kW) yielded dried ginseng of relatively high porosity (41.14 %), good appearance and bioactive compounds content slightly higher (~33 mg ginsenosides/g dry ginseng) than the freeze-dried product, with the advantage of presenting a shorter drying time.

Yan et al. [43] compared the effect of different microwave-assisted drying processes on quality and engineering aspects of carrots. Samples were obtained by dicing and blanching the carrots in hot water. Microwave-enhanced spouted bed dryer was operated with forced air at 50 °C and microwave power density of 2.0 W/g or 3.5 W/g (w.b.). Microwave-assisted vacuum dryer was operated at an absolute pressure of 5 kPa, microwave power density equal to 2.4 W/g (w.b.) and turntable rotation speed of 5 rpm. Microwave-assisted freeze-dryer (Fig.2) was operated at microwave power density of 2.0 W/g, absolute pressure of 100 Pa and condenser (cold trap) temperature of −40 °C. Samples were evaluated for moisture, β-carotene and vitamin C contents, along with rehydration ratio, color, and sensory characteristics. The drying processes were assessed for energy consumption expressed as power consumption per kilogram of evaporated water. All samples were dried to 8 % of moisture (w.b.). It was observed that the highest drying rates were yielded by the microwave-enhanced spouted bed drying at higher power level, while the lowest drying rates were yielded by the microwave-assisted freeze-drying. As regards rehydration capacity, microwave freeze-dried product presented the best behavior, being very similar to the reference sample that was conventionally freeze-dried. On the other hand, the other tested methods yielded products of rehydration capacity similar to each other and significantly inferior to the microwave freeze-dried product. Dried carrot color was less affected by microwave freeze-drying and microwave-vacuum drying than by microwave-enhanced spouted bed in terms of L*, a* and b* changes. Nevertheless, microwave-enhanced spouted bed drying generated no charred samples, while microwave freeze-drying and microwave-vacuum drying yielded roughly 1 and 6 % of charred samples, respectively. β-carotene and vitamin C were significantly better preserved in the microwave freeze-dried product as compared to the other methods. Sensory scores showed that microwave freeze-dried carrot presented better appearance and texture, while color and flavor scores were statistically similar for all drying methods. In general terms, all samples were well accepted by the sensory panel. Finally, energy consumption was significantly higher for microwave freeze-drying when compared to the other methods. Concluding, microwave freeze-drying presented the same advantages and disadvantages of regular freeze-drying, i.e., high product quality and low energy efficiency. Microwave-vacuum drying and microwave-enhanced spouted bed drying presented advantages concerning energy aspects and strengths/weaknesses regarding product quality.

Antal et al. [3] investigated the effect of hot air drying and freeze-drying on moisture content, volatile compounds and rehydration ratio of spearmint leaves. Drying conditions were: air temperature of 43 °C, air relative humidity of 11–48 % and air velocity of 0.5 m/s, for hot air drying; and 17 °C heating plate temperature, −55 °C condenser temperature and pressure of 10–30 Pa or 150–250 Pa. The drying curves were tentatively fitted with the Lewis, the Page, the third degree polynomial and the sigmoid models. It was observed that hot air drying yielded higher rates than freeze-drying. Additionally, pressure reduction led to higher freeze-drying rates. A drying-rate curve showed that freeze-drying started at low rates, reached a maximum rate approximately in the middle of the process and then decreased. All tested models provided a proper fit for the drying kinetics. Regarding the concentration of volatile compounds in the dried product, hot air drying led to extensive destruction of volatiles. Freeze-drying preserved very well the volatiles, especially under high pressures. It was demonstrated that high vacuum helped the essential oils to escape from the leaves. Rehydration of dried leaves in water was shown to be very fast at the beginning and slow afterwards. Rehydration curves were well fitted by a two-term exponential model. Freeze-dried products rehydrated much better than hot air dried ones. Furthermore, increase in rehydration water temperature from 35 to 75 °C improved rehydration. In conclusion, freeze-drying at not too low pressure was recommended by those authors for yielding high quality dried spearmint leaves.

Huang et al. [16] investigated how different methods of drying affected the quality of apple/potato re-structured chips. The drying methods used were microwave-freeze drying (MFD), freeze-drying (FD), microwave-vacuum drying (MVD) and vacuum drying (VD). Process conditions were as follows: 100 Pa, −40 °C condenser temperature, 50 °C product surface temperature and 1.6 W/g of power density for MFD; 100 Pa, −40 °C condenser temperature and 50 °C heating plate temperature for FD; microwave power density of 4 W/g and pressure of 25 kPa until reaching 40 % product moisture content, decreasing to 5 kPa until the end of drying for MVD; 50 °C and the same pressure combination used for MVD, for VD. Chips of 3 mm thickness were dried until 6 % (w.b.) moisture content was reached. Pre-tests showed that the optimum potato to apple ratio was 65:35 (w/w). Drying curves showed that MVD yield the highest drying rates, followed by MFD, FD and VD. Texture analysis showed that the crispiest chips are those produced by VMD, while the toughest are those produced by VD. As for bulk density, FD samples were the best, i.e., lowest bulk density, while VD samples were the worst, i.e., densest. Vitamin C was more preserved in MFD samples and more destructed in VD samples. As regards color, MFD and FD chips were lighter than MVD and VD chips. On the other hand, MVD and VD chips were desirably yellower. VD chips, though, turned brown as denoted by positive a* values. A sensory evaluation demonstrated that MFD chips present the best quality, followed by FD, MVD and VD. Nevertheless, VD chips did not achieve the minimum acceptable score. MFD chips presented the best rehydration, followed by FD, VMD and VD. Micrographs helped to explain the rehydration results: even though the chips were made from pastes, freezing prior to MFD and FD led to the formation of ice crystals that let a porous structure when sublimated. This structure was clearly honeycomb-like in MFD chips, which consequently rehydrated better. In conclusion, MFD yielded chips of high quality, being recommended for yielding premium products, since it is associated with high costs. On the other hand, MVD yielded chips of acceptable quality at a low cost, being recommended for large scale production.

Vashisth et al. [39] studied the influence of different conditions of hot air drying, freeze-drying and vacuum belt drying on the moisture, water activity, phenolics content and antioxidant activity of muscadine grape pomace. Drying procedures were conducted under the following conditions: hot air drying at 70–80 °C, 0.2–0.6 m/s air velocity and 180–240 min; freeze-drying at 0.5 Pa, heating plate temperature of 30 °C and 14–16 h; vacuum belt drying at 60–90 °C in the first zone of the dryer, 80–105 °C in the second zone of the dryer, 100–120 °C in the third zone of the dryer and 100–120 °C in the fourth zone of the dryer, total residence times of 60–90 min and pressure varying from 3–5 kPa. An example of equipment for performing vacuum belt drying will be presented in this chapter. Product thickness was either 2 or 4 mm. Results showed that 2 mm samples were dried at higher rates than 4 mm samples. The relationship between moisture content and water activity followed an exponential pattern for all treatments and thicknesses. Phenolics content was statistically similar for freeze-dried and vacuum belt dried products, while hot air dried products presented lower values. In addition, phenolics were more preserved in 4 mm samples as compared to 2 mm samples. With regard to antioxidant activity as measured by the FRAP method, results varied a lot, with some conditions of hot air drying yielding products with similar antioxidant activity of freeze-dried and vacuum belt dried samples. In sum, vacuum belt drying showed proper for yielding good quality product in a feasible drying time. As expected, the freeze-dried product presented proper quality, but the high drying times associated with high costs impairs the use of this technique in practical terms.

Liu et al. [22] studied the influence of microwave-assisted vacuum drying (MWVD), microwave-assisted freeze-drying (MWFD) and microwave-assisted spouted bed drying (MWSBD) on quality and energy aspects of purple-fleshed sweet potato. Such vegetable presents a healthy appeal due to its high anthocyanins content. Initially, the purple-fleshed sweet potatoes were peeled, sliced at 10 mm and steamed for 10 min. Then, they were blended with other ingredients and shaped into cylindrical granules before drying. The drying experiments were performed in the following way: MWVD at a 4 W/g power density, a 4.5 kPa absolute pressure and a 10 rpm plate rotational speed; MWFD at a 100 Pa absolute pressure, a 45 °C heating plate temperature and a condenser temperature of −38 °C; and MWSBD at an air temperature of 80 °C and a power density of 2.5 W/g. The product was evaluated for moisture content, color, anthocyanin level, texture and sensory quality. The process was evaluated for energy consumption. The products were dried to <6 % (w.b.) final moisture content. Results showed that the faster process was MWSBD, followed closely by MWVD and much faster (nine times) than MWFD. The original color was better preserved by MWFD and MWVD, while MWSBD yielded a more burnt and less purple product. Such results were confirmed by the anthocyanins quantification which showed lower levels of these pigments for the MWSBD sample. Texture instrumental analysis showed that the hardest product was that processed by MWVD, while the softest was the one processed by MWSBD. Sensory quality was expressed in terms of appearance, crispness, color and flavor. MWSBD product was found with better appearance and color, while MWFD one was found with better crispness and flavor. Total scores were higher for MWSBD product, followed by MWFD and MWVD, the latter not reaching the minimum level of acceptability. Finally, energy consumption was higher for MWFD, intermediate for MWVD and lower for MWSBD. Therefore, those authors concluded that MWSBD is a feasible way to obtain purple-fleshed sweet potato products, these presenting a quality as high as those yielded by MWFD with the advantage of consuming much less energy. Nevertheless, they observed that some kind of process control must be put in place in order to avoid extensive destruction of anthocyanins during MWSBD.

Sansiribhan et al. [33] tried to validate the use of indicators of microstructural changes for expressing the structure-quality of carrot cubes during different drying methods. Drying was performed by: low pressure superheated steam drying (LPSSD) at 60–80 °C and pressure of 7 kPa; vacuum drying, under the same conditions used for LPSSD, but without the injection of vapor; and hot air drying at 60–80 °C and a 0.8 m/s air velocity. The microstructural changes were expressed in terms of normalized change fractal dimension and normalized change of cell diameter. Shrinkage and hardness were used as structure-quality indicators. The drying curves were also generated. The carrots were dried until a 0.1 kg/kg (d.b.) moisture content was achieved. When comparing the drying curves generated for

Fig. 3 Schematic of a continuous vacuum belt dryer [41]

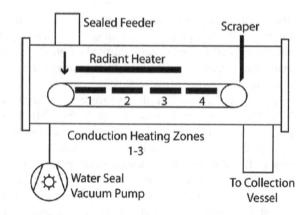

the three processes, it was observed that vacuum drying was faster than LPSSD and hot air drying. Higher drying temperatures led to lower drying times, as expected. Changes in fractal dimension and changes in cell diameter increased with the advance of drying. Such relationships were well explained by logarithmic empirical models. When correlating macro- and microstructural properties, it was shown that: the higher the shrinkage, the higher the changes in fractal dimension and cell diameter; and the higher the hardness, the higher the changes in fractal dimension and cell diameter. The relationships between microstructural changes and shrinkage changes followed an exponential pattern. On the other hand, the relationships between microstructural changes and hardness changes followed a logarithmic pattern. All correlations yielded high values of correlation coefficient, even though lower values ($r = 0.851$) were obtained for the correlation between changes in cell diameter and shrinkage. Therefore, those authors concluded that the microstructural parameters tested, namely normalized changes in fractal dimension and normalized changes in cell diameter, are valid for reflecting the structure-quality changes of carrots during drying under various processes and conditions, the former being considered better than the latter.

Xu and Kerr [41] compared the effect of continuous vacuum drying (CVD) and deep fat frying (DFF) on quality characteristics of corn chips. The product comprised cooked corn/salt/water dough sheeted to 1 or 3 mm and cut into circles with a 7 cm diameter. CVD was performed in a pilot scale continuous vacuum belt dryer (Fig. 3) set to a 90 °C plate temperature and a 3,000 Pa absolute pressure. The CVD process lasted 75 min. DFF was performed in a stainless fryer at 180 °C. The DFF process lasted from 45 to 120 s, depending on product thickness. The final product was evaluated for oil content, sensory attributes, color and texture. Results showed that the oil content of CVD chips ranged from 1.57 to 1.82 g oil/100 g, while DFF chips presented from 33.37 to 34.80 g oil/100 g. Color was better preserved in CVD chips, with higher lightness (L*), higher colorfulness (C*) and a hue closer to pure yellow as compared to DFF chips. A previous sensory analysis showed that both products presented a good purchase intention. In addition, fat content was a parameter taken into account by 71 % of the panelists when buying a snack. Nevertheless, a subsequent sensory evaluation showed that DFF chips were slightly more liked than CVD chips. In fact, the most liked product was the 1 mm thick DFF chip. Crispness is an appreciated attribute in snacks, which made those authors perform a comprehensive texture analysis in their chips. Fracture force, i.e., the maximum force required to break the sample, was higher in CVD chips. The texture evaluation also comprised an acoustic analysis. In this regard, CVD chips presented a higher number of sound peaks, higher total sound energy and higher percentage of sounds in the frequency below 2,500 Hz when compared to DFF chips. The meaning of these results is that, according to the acoustic analysis, CVD chips were crispest than DFF chips. Nevertheless, the sensory evaluation showed that DFF chips were more appreciated than CVD chips as regards texture, possibly because crispness is not the only attribute determining acceptability. Summarizing, continuous vacuum drying was able to yield corn chips with good consumer acceptability, probably due to a preserved color and a crispy texture. The greatest advantage of the vacuum dried corn chips is their low fat content when compared to deep fat fried products.

The effect of various drying techniques on the composition of thyme essential oil and other quality parameters was studied by Calín-Sánchez et al. [4]. Thyme plants were dried either by: convective drying (CD) at 40–60 °C and air velocity of 0.8 m/s; vacuum-microwave drying (VMD) at 4–6 kPa absolute pressure and 240–480 W microwave power level; convective predrying followed by vacuum-microwave finish drying (CPD-VMFD) with CPD being performed at 40–60 °C until moisture content of 3 kg/kg (d.b.) followed by VMFD at 240–480 W; and freeze-drying (FD) at a 65 Pa absolute pressure, a −40 °C condenser temperature and a 30 °C heating plate temperature. Results showed that CD kinetics was well described by a two-term exponential model. VMD kinetics was well described by a linear function (constant rate period) in combination with an exponential function (falling rate period). As expected, the use of higher air temperatures and high power levels led to higher drying rates. With concern to the volatiles composition of the fresh thyme samples, results showed that the most abundant compound was thymol, followed by γ-terpinene, p-cymene, caryophyllene and α-terpinene. Therefore, these are the main responsible for the thyme aroma. With concern to the volatiles composition of the dried thyme, CD caused the highest volatiles loss, while VMD preserved the volatiles in the best way. The hybrid method also yielded promising results, especially when CPD was followed by VMD at 240 W. The loss of volatile compounds was enhanced by higher air temperature and higher microwave power level. A sensory descriptive analysis yielded results that, in general terms, support those obtained for the volatile compounds analysis. In other words, the assessors perceived an aroma close to that of fresh thyme in samples dried by VMD at low power level and in samples dried by CPD-VMFD at low air temperatures and low power levels. To summarize, those authors recommended using CPD-VMFD at 40 °C and 240 W for obtaining dried thyme with the best aromatic properties.

The influence of osmotic dehydration (OD) time on quality aspects of hot air (HA) and microwave-vacuum (MVD) dried papaya cubes was studied by Nimmanpipug et al. [26]. Papayas are appreciated all around the world and drying is a way to provide them with unique sensory characteristics and extend their shelf-life. HA was compared to MVD in terms of product color, texture, rehydration capacity, glass transition temperature (T_g) and sensory acceptance. OD was carried out in a 65 % (w/w) sucrose aqueous solution at 40 °C and 40 rpm agitation for 1–4 h. HA was conducted at a 70 °C air temperature and a 1 m/s air velocity. MVD was performed at a 3.75 W/g power intensity and a 13.33 kPa pressure. Effective moisture diffusivities were calculated by using Fick's second law of diffusion (Eq. 1 in chapter Studies on Conventional Vacuum Drying of Foods). All products were dried until their moisture content reached <0.2 kg/kg (d.b.). Results indicated that OD reduced the papayas moisture content to about 4.51, 2.36, 1.53 and 1.22 kg/kg (d.b.) after 1, 2, 3 and 4 h of process, respectively. In fact, the three first hours exerted more effect on product moisture content. Nevertheless, increase in OD duration promoted a decrease in effective moisture diffusivity (D_{eff}). D_{eff} values ranged from 7.09 to 9.13 $\times$ 10^{-8} m^2 s^{-1} for HA and from 2.85 to 3.50 $\times$ 10^{-6} m^2 s^{-1} for MVD. It was found that HA dried papaya color was closer to red while MVD dried papaya color was closer to orange. Drying by both methods decreased the lightness of OD dehydrated papaya, probably due to degradation of thermo-labile pigments and browning reactions. MVD dried samples were lighter than HA dried ones, yet less saturated (lower C*). The use of OD increased T_g, leading to more stable products. Indeed, the longer the OD, the higher the T_g value. Regarding texture, HA dried product was harder than MVD dried one, probably due to shrinkage and case hardening. A hardening effect was observed during both processes, being this effect attenuated by OD. A decrease in product gumminess was also observed in OD dehydrated papaya. Both drying methods caused decrease in springiness, i.e., elasticity. Regarding rehydration, the longer the OD, the better and faster was the dried papaya rehydration. Comparing both methods, the MVD dried papaya rehydrated better than the HA dried one. Finally, a Principal Component Analysis (PCA) indicated that high values of water activity, lightness, hue angle and springiness are related to high quality dried papaya. On the other hand, high values of hardness, gumminess, T_g and saturation are related to poor quality dried papaya. In general terms, MVD preceded by OD during 1–3 h provided the best quality dried papaya.

   Wang et al. [40] compared the effects of vacuum drying (VD), microwave-vacuum drying (MVD) and the innovative pulsed spouted microwave-vacuum drying (PSMVD) on quality parameters of stem lettuce slices. The analyses of drying uniformity and some drying characteristics were carried out only for MVD and PSMVD. On the other hand, the dried products quality assessment involved the VD, the MVD and the PSMVD dried products. PSMVD was performed at a 20 °C air temperature, a 30 % air relative humidity, a 3.5 m/s air velocity, a 7–10 kPa absolute pressure and a 480 W power level. MVD was performed at a 7 kPa absolute pressure and a 480 W power level. The magnetrons were operated at 2,450 MHz. VD was carried out at a 7 kPa absolute pressure and a 60 °C temperature. All samples were dried to 6.5 % final moisture content on wet basis. Results showed that, for drying the same 200 g of sample, PSMVD took 60 min while MVD took roughly 120 min. Product temperature curves showed that the temperature of the slices positioned at the center and at the periphery of the chamber increased on initial stages and reached a plateau at the end, differently from what was observed in previous studies. The temperature distribution analysis showed that the MVD dried samples positioned at the center of the turntable were much different from those located at the periphery. This was not observed in PSMVD. Product moisture content as measured in twelve different slices varied significantly from one slice to another during MVD, with central slices containing more moisture than peripheral slices. Such effect was much attenuated during PSMVD. Color of the MVD dried product was also less uniform as compared to the PSMVD dried one. When comparing color of the fresh, PSMVD, MVD and VD dried products, lightness increased in this exact order. Negative values of $a*$ and positive values of $b*$ indicated that the product was green and yellow. Regarding total color difference from the fresh product, the best results were provided by PSMVD, followed by MVD. With concern to shrinkage, PSMVD and MVD yielded similar results. The microstructure of stem lettuce slices as dried by PSMVD, MVD and VD was compared to that of freeze-dried samples. It was observed that the three studied methods promoted more shrinkage than freeze-drying, as denoted by tightly linked cells. Apparent relative density was higher for the PSMVD dried product and lower for MVD dried products. Rehydration capacity was higher for the VD dried products and lower for the MVD dried ones. Rehydrated product hardness was higher for the PSMVD dried samples, implying a more elastic behavior, while lower values of this parameter were observed for the MVD dried samples, implying a softer texture. Concluding, the new process named pulsed spouted microwave-vacuum drying presents many advantages over the existing techniques in terms of drying time, drying uniformity and product quality.

The drying of vacuum-impregnated apples was comparatively studied by Schulze et al. [34]. Apples were peeled and sliced before all treatments. Then, a vacuum impregnation process was performed by immersion of the slices in a flavonoids-enriched apple juice contained in a recipient placed inside a vacuum chamber at 10 kPa for 5 min at room temperature. This procedure was followed by drying by either: microwave-vacuum drying at 2 kPa and the power levels/time of 500 W/25 min, followed by 0 W/5 min, then 1,000 W/1 min and finally 80 W/until the end of the process (total drying time of 130 min); freeze-drying at undefined process conditions for 72 h; and hot air drying at 50 °C for 14 h. It was observed that hot air dried apples presented high browning and high shrinkage, while freeze-dried ones presented low shrinkage and low browning. Microwave-vacuum dried samples presented high shrinkage but low browning. Freeze-dried samples presented the lowest moisture content and water activity, while hot air dried ones presented the highest values of these parameters. With regard to the preservation of bioactive compounds, i.e. quercetin derivatives, freeze-drying and microwave-vacuum drying led roughly to the same results, while hot air drying destructed more extensively these compounds. Color of hot air dried samples was much impaired, while the other dried apples presented a more natural color. Bulk density was high for hot air- and microwave-vacuum dried apples, while freeze-dried ones presented low bulk density. Finally, storage at 20 °C for 12 months did not affect significantly the product quercetin derivative contents. Summarizing, microwave-vacuum drying provided an interesting balance between drying time and product quality, while freeze-drying presented the drawback of elevated drying time and hot air drying presented the disadvantage of inferior product quality.

References

1. Acevedo NC, Briones V, Buera P et al (2008) Microstructure affects the rate of chemical, physical, and color changes during storage of dried apple discs. J Food Eng 85:222–231
2. Alibas I (2009) Microwave, vacuum, and air drying characteristics of collard leaves. Dry Technol 27:1266–1273
3. Antal T, Figiel A, Kerekes B et al (2011) Effect of drying methods on the quality of the essential oil of spearmint leaves (*Mentha spicata* L.). Dry Technol 29:1836–1844
4. Calín-Sánchez A, Figiel A, Lech K et al (2013) Effects of drying methods on the composition of thyme (*Thymus vulgaris* L.) essential oil. Dry Technol 31:224–235
5. Caro-Corrales JJ, Zazueta-Niebla JA, Ordorica-Falomir CA et al (2005) Controlled low-temperature vacuum dehydration and tunnel drying: a comparative study. Int J Food Prop 8:529–542
6. Cui Z-W, Xu SY, Sun D-W (2003) Dehydration of garlic slices by combined microwave-vacuum and air drying. Dry Technol 7:1173–1184
7. Cui Z-W, Xu S-Y, Sun D-W (2004) Effect of microwave-vacuum drying on the carotenoids retention of carrot slices and chlorophyll retention of Chinese chive leaves. Dry Technol 22:563–575
8. Deng Y, Zhao Y (2008) Effect of pulsed vacuum and ultrasound osmopretreatments on glass transition temperature, texture, microstructure and calcium penetration of dried apples (Fuji). LWT-Food Sci Technol 41:1575–1585
9. Devahastin S, Suvarnakuta P, Soponronnarit S et al (2004) A comparative study of low-pressure superheated steam and vacuum drying of a heat-sensitive material. Dry Technol 22:1845–1867
10. Dunlap WC Jr (1946) Vacuum drying of compressed vegetable blocks. Ind Eng Chem 38:1250–1253
11. Giri SK, Prasad S (2007) Drying kinetics and rehydration characteristics of microwave-vacuum and convective hot-air dried mushrooms. J Food Eng 78:512–521
12. Grabowski S, Marcotte M, Poirier M et al (2002) Drying characteristics of osmotically pretreated cranberries: energy and quality aspects. Dry Technol 20:1989–2004
13. Hawlader MNA, Perera CO, Tian M et al (2006) Drying of guava and papaya: impact of different drying methods. Dry Technol 24:77–87
14. Hiranvarachat B, Suvarnakuta P, Devahastin S (2008) Isomerisation kinetics and antioxidant activities of β-carotene in carrots undergoing different drying techniques and conditions. Food Chem 107:1538–1546
15. Hossain MB, Barry-Ryan C, Martin-Diana AB et al (2010) Effect of drying method on the antioxidant capacity of six Lamiaceae herbs. Food Chem 123:85–91
16. Huang L-I, Zhang M, Mujumdar AS et al (2011) Comparison of four drying methods for restructured mixed potato with apple chips. J Food Eng 103:279–284
17. King VAE, Zall RR, Ludington DC (1989) Controlled low-temperature vacuum dehydration—a new approach for low-temperature and low-pressure food drying. J Food Sci 54:1573–1579
18. Krokida MK, Kiranoudis CT, Maroulis ZB (1999) Viscoelastic behavior of dehydrated products during rehydration. J Food Eng 40:269–277

19. Krokida MK, Maroulis ZB, Saravacos GD (2001) The effect of the method of drying on the colour of dehydrated products. Int J Food Sci Tech 36:53–56

20. Li Y, Xu S-Y, Sun D-W (2007) Preparation of garlic powder with high allicin content by using combined microwave-vacuum and vacuum drying as well as microencapsulation. J Food Eng 83:76–83

21. Lin TM, Durance TD, Scaman CH (1998) Characterization of vacuum microwave, air and freeze dried carrot slices. Food Res Int 31:111–117

22. Liu P, Zhang M, Mujumdar AS (2012) Comparison of three microwave-assisted drying methods on the physiochemical, nutritional and sensory qualities of re-structured purple-fleshed sweet potato granules. Int J Food Sci Tech 47:141–147

23. Maache-Rezzoug Z, Rezzoug SA, Allaf K (2001) Kinetics of drying and hydration of the scleroglucan polymer. A comparative study of two conventional drying methods with a new drying process: dehydration by successive pressure drops. Dry Technol 19:1961–1974

24. Maroulis ZB, Tsami E, Marinos-Kouris D et al (1988) Application of the GAB model to the sorption isotherms of dried fruits. J Food Sci 7:63–78

25. Martínez-Soto G, Ocaña-Camacho R, Paredes-López O (2001) Effect of pretreatment and drying on the quality of oyster mushrooms (*Pleurotus ostreatus*). Dry Technol 19:661–672

26. Nimmanpipug N, Therdthai N, Dhamvithee P (2013) Characterization of osmotically dehydrated papaya with further hot air drying and microwave vacuum drying. Int J Food Sci Tech 48:1193–1200

27. Panyawong S, Devahastin S (2007) Determination of deformation of a food product undergoing different drying methods and conditions via evolution of a shape factor. J Food Eng 78:151–161

28. Purnama M, Yaghmaee P, Durance TD et al (2010) Porosity changes and retention of ginsenosides in North American ginseng root using different dehydration processes. J Food Sci 75:E487–E492

29. Qing-guo H, Min Z, Mujumdar AS et al (2006) Effect of different drying methods on the quality changes of granular edamame. Dry Technol 24:1025–1032

30. Rahman MS, Al-Amri OS, Al-Bulushi IM (2002) Pores and physico-chemical characteristics of dried tuna produced by different methods of drying. J Food Eng 53:301–313

31. Rahman MS, Al-Shamsi QH, Bengtsson GB et al (2009) Drying kinetics and allicin potential in garlic slices during different methods of drying. Dry Technol 27:467–477

32. Rodríguez R, Lombraña JI, Kamel M et al (2005) Kinetic and quality study of mushroom drying under microwave and vacuum. Dry Technol 23:2197–2213

33. Sansiribhan S, Devahastin S, Soponronnarit S (2012) Generalized microstructural change and structure-quality indicators of a food product undergoing different drying methods and conditions. J Food Eng 109:148–154

34. Schulze B, Hubbermann EM, Schwarz K (2014) Stability of quercetin derivatives in vacuum impregnated apple slices after drying (microwave vacuum drying, air drying, freeze-drying) and storage. LWT-Food Sci Technol 57:426–433

35. Sunjka PS, Rennie TJ, Beaudry C et al (2004) Microwave-convective and microwave-vacuum drying of cranberries: a comparative study. Dry Technol 22:1217–1231

36. Suvarnakuta P, Devahastin S, Mujumdar AS (2005) Drying kinetics and β-carotene degradation in carrot undergoing different drying processes. J Food Sci 70:S520–S526

37. Thomkapanich O, Suvarnakuta P, Devahastin S (2007) Study of intermittent low-pressure superheated steam and vacuum drying of a heat-sensitive material. Dry Technol 25:205–223

38. Tsami E, Krokida MK, Drouzas AE (1999) Effect of drying method on the sorption characteristics of model fruit powders. J Food Eng 38:381–392

39. Vashisth T, Singh RK, Pegg RB (2011) Effect of drying on the phenolics content and antioxidant activity of muscadine pomace. LWT-Food Sci Technol 44:1649–1657

40. Wang Y, Zhang M, Mujumdar AS (2013) Study of drying uniformity in pulsed spouted microwave-vacuum drying of stem lettuce slices with regard to product quality. Dry Technol 31:91–101

41. Xu S, Kerr WL (2012) Comparative study of physical and sensory properties of corn chips made by continuous vacuum drying and deep fat frying. LWT-Food Sci Technol 48:96–101

42. Yaghmaee P, Durance T (2007) Efficacy of vacuum microwave drying in microbial decontamination of dried vegetables. Dry Technol 25:1109–1114
43. Yan W-Q, Zhang M, Huang L-L et al (2010) Studies on different combined microwave drying of carrots pieces. Int J Food Sci Tech 45:2141–2148
44. Yang W-C (2003) Particle characterization and dynamics. In: Yang W-C (ed) Handbook of fluidization and fluid particle system. Marcel Dekker, New York, pp 1–24
45. Yanyang X, Min Z, Mujumdar AS et al (2004) Studies on hot air drying and microwave vacuum drying of wild cabbage. Dry Technol 22:2201–2209